IMAGINATION AND
A PILE OF JUNK

Trevor Norton is an Emeritus Professor at the University of Liverpool, having retired from the Chair of Marine Biology. He has published widely on ecological topics. He is also an Honorary Senior Fellow at the Centre for Manx Studies on the Isle of Man where he lives.

His much acclaimed books include *Stars Beneath the Sea*, *Reflections on a Summer Sea*, and *Under Water to Get Out of the Rain*.

IMAGI-NATION

A droll history of inventors and inventions

Trevor Norton

CORONET

First published in Great Britain in 2014 by Coronet
An imprint of Hodder & Stoughton
An Hachette UK company

First published in paperback in 2015

1

Copyright © Trevor Norton 2014

ISBN 978 1 444 73258 0

Printed and bound by Clays Ltd, St Ives plc

Hodder & Stoughton policy is to use papers that are natural,
renewable and recyclable products and made from wood grown in
sustainable forests. The logging and manufacturing processes are
expected to conform to the environmental regulations of the country
of origin.

Hodder & Stoughton Ltd
338 Euston Road
London NW1 3BH

www.hodder.co.uk

PREFACE

It was Thomas Edison who claimed that junk and imagination were all that the inventor needed. He omitted to mention that he also had two hundred assistants.

The traditional view of an inventor is Archimedes leaping half-washed from his bath and running naked down the high street shouting, 'Eureka – I've got it!' What he'd got was surely plain for all to see. His joy probably diminished when having to explain his behaviour to the magistrate.

Real-life inventors are a mixed bunch of bold, brilliant and sometimes barmy eccentrics. In their obsessive pursuit of something new they may neglect bathing altogether. The rare eureka moment was often followed by years of toil. James Dyson made five thousand prototypes over fifteen years before perfecting his ingenious vacuum cleaner. They succeeded because of their ingenuity and doggedness or luck and a flair for 'adopting' and adapting other inventors' devices.

Technology has shaped both the ancient and modern world and now we are entirely reliant on devices. Sober citizens would be reduced to a weeping jelly if we confiscated their smartphones and laptops.

Almost all advances are taken for granted. Without vaccines half my readers would have been dead before they were old enough to read. We have completely forgotten the self-doffing hat and the stylish cloak that transformed into a dinghy with a paddle in the pocket.

Brace yourself for things weird and wonderful – inventors made them all.

To my wonderful wife Win and
the greatest inventions of all, my grandchildren:
Charlotte, Katie and Matthew,
without whom this book would have
been written in half the time

ACKNOWLEDGEMENTS

I am grateful to the Director of the Centre for Manx Studies for tolerating 'the man who orders weird books'. I have benefited greatly from access to the libraries of Liverpool University and the medical library of Keyll Darree, the DHSS Education & Training Centre in the Isle of Man. My thanks also extend to Paul Ogden and the Trustees of the Milntown Estate for access to the excellent library on vintage automobiles.

I am indebted to those who facilitated my access to information: Dr Andrew Brand, Angela and her staff at the Bridge Bookshop in Port Erin, Isle of Man, and Drs Terry and Selma Holt.

Sincere thanks also to Rachel Norton Buchleitner and Richard Dawes for their meticulous proofreading, Anna Webber at United Agents, and Mark Booth at Hodder & Stoughton who commissioned the book.

CONTENTS

INVENTIONS BEFORE THERE WERE INVENTORS

'Countless things that humanity acquired in earlier stages
. . . suddenly emerge into the light'
– Nietzsche

S urely it requires someone smart to invent things – someone
like us. *Homo sapiens* means 'wise man' in Latin, although
sometimes I think it derives from 'sap', American slang for
a gullible fool. Modern man has been around for only 200,000
years or so in a universe 13.7 billion years old and in that
time he has achieved so much.

Developing even simple technologies took a long time and
not all of the early breakthroughs were made by our species.
Long ago when human beings were just a gleam in evolution's
eye, creatures roamed the earth that were no longer apes but
not yet human. Africa seems to have spawned a variety of
ape men. There were, of course, ape *women* too, and indeed
we can all trace our ancestry back to a single female called
'mitochondrial Eve'. But 'ape person' or 'Neanderthal woman'
just doesn't sound right, so I will use 'man' to refer to the
species, rather than the gender that invented alcohol and
falling over.

Many of these early ape men existed millions of years ago

and the first specimens found were given cosy names such as 'The Hobbit' or 'Ida' (after the discoverer's five-year-old daughter) or 'Lucy' (from the Beatles' 'Lucy in the Sky with Diamonds'). The more distinguished-sounding 'Proconsul' was named after a performing monkey at the *Folies Bergère*.

We know our earliest antecedents solely from their bones. Then, around 2.6 million years ago, they began to leave other calling cards – pieces of worked stone. Ape men were slow learners. They had been around for eleven million years before they fashioned their first tool. A skeleton three million years old was found in Africa surrounded by crudely shaped tools and a litter of the rock fragments chipped away during their manufacture. The tool makers were species called *Homo ergaster* (work man) and *Homo habilis* (skilled man). As many women know, these types of men are now extinct.

Necessity was always the mother of invention and nothing is more necessary than survival. Tools were a great aid to feeding and fighting. Some types of rock could be struck and sharp-edged pieces would break off. Fossilised bones of animals are scarred where the meat was cut and scraped off with just such tools. A simple form of fish hook was whittled from wood or bone to ensure that fish were part of their diet. Two fatty acids scarce in land animals but abundant in aquatic creatures promoted brain growth and the early ape men needed all the brain expansion they could get.

The first weapon to be invented was the hand axe, which was not really an axe at all but a multi-purpose tool that could kill, cut, skin and scrape. It was a rock roughly rounded at one end to fit into the hand and shaped into a point at the business end.

Then along came *Homo erectus* – the name describes their

posture, not a propensity for arousal. Their brain was thirty per cent smaller than ours, yet they manufactured flint tools with razor-sharp edges. They carried knives and hunted in packs just like present-day gangs.

Many archaeologists think that *Homo erectus* tamed fire well over a million years ago. There are finds of charred bones and charcoal from that date, so they may have used fire but had they devised how to *make* fire? Fire-igniting sparks can be made striking flint against iron pyrites (fool's gold) and the first evidence of that dates from 790,000 years ago. Wild fires were terrifying, so it took a bold fellow to bring fire into the home. Fire lighting may have been discovered by accident. Perhaps a tool maker chipping away at a lump of flint generated a spark which ignited the dry litter on the cave floor. If the consequent conflagration didn't incinerate all his possessions and send his family scurrying for their lives, he might have realised he had invented something wonderful, as long as it could be contained.

Fire was important in our evolution. A hearth gave warmth in severe weather and light during the night, as well as keeping nocturnal predators at bay. Most important of all, it enabled food to be cooked. Raw food is hard work to chew and digest, but when cooked it slips down easily. A taste for cooked food may have arisen after a joint of meat accidentally fell into the fire or as a result of scavenging roasted carcasses in the aftermath of a wild fire. The invention of roasting and boiling food would rapidly follow.

This was not merely a culinary advance. Meat is rich in calories and cooking released more nutrition to feed the brain and fuel our developing intellect. Over two million years the skulls of ape men quadrupled in size. In addition a cooking ape devoted far less time to chewing. Simians spend half their waking hours chomping their food. That time and energy could be better spent hunting, making tools or socialising.

Homo sapiens evolved in Africa and Asia, then dispersed, arriving in Europe about forty-five thousand years ago. By that time Neanderthals (named after the Neander Valley in Germany where the first specimen was found) had been the dominant ape men in Europe for 200,000 years. Neanderthal man has had a bad press. Yes, he was built like the Incredible Hulk and the overhang of his brow would challenge even a skilled rock climber. He was not someone you would wish to meet on a dark night or even in full daylight. But he wasn't a hairy, grunting dullard whose idea of foreplay was to bash a girl over the head with his club.

The Neanderthals were sufficiently human-looking that when some skeletons were discovered in Spain in 1994, the police were called. Their brain case was slightly *larger* than ours and they had the same vocal apparatus, so, although not great conversationalists, they probably communicated using a rudimentary language. They were smart enough to fashion tools that fitted the hand and spears that had fire-hardened points. Neanderthals were also the first ape men to bury their dead, leave funerary offerings and care for the sick and injured, of which there were many.

Neanderthal skeletons often had fractured or crushed bones, which is attributed to their hunting technique. Although they invented heavy-duty spears and knives, these were close-proximity jabbing weapons. The hunter relied on brute strength to bring an animal down, and whether the man or his prey survived was in the balance.

In comparison the newly arrived *Homo sapiens* were

puny-looking specimens, slim and with a slim chance of surviving in a landscape swarming with hyenas, woolly rhinos and big cats. Surely they would be no match for either the wildlife or the Neanderthals, yet it was they who prevailed. These modern men had long, lightweight spears with sharp flakes of flint embedded in their tips and they used them as projectiles. They also invented the spear-thrower, a device that launched the lance with greater force and velocity. One of these spears was found lodged in the skeleton of a huge mastodon. They could weaken or kill even large animals from a distance and with less risk. The technology available to *Homo sapiens* included a wide range of specialised tools honed from flint, bone and antler. The advent of the sling shot and bow and arrow marked the beginning of man's dominance over all other creatures.

It was a period of climatic fluctuation between icy and subtropical periods. The Neanderthals were disadvantaged by their physiology: their daily calorific requirement was almost double that of *Homo sapiens*, a great liability when food was scarce. Unlike the Neanderthals, our ancestors invented needles and were able to sew animal hides together to make warm, windproof clothing.

By thirty thousand years ago *Homo sapiens* were manufacturing jewellery, carving statuettes of super-voluptuous women and painting caves with stunning dioramas of wild animals and hunting scenes. They had invented art and decoration, which later became symbols of culture. It is ironic that they found time to spend on non-essentials shortly after the Neanderthals lost the struggle for survival. Our ancestors outlived all the other species of humanoid apes.

But the Neanderthals left a faint echo in our DNA. Clearly,

while the two species overlapped, some individuals took the overlapping too literally. Consequently if you are European, not of African origin, then one to four per cent of your genetic makeup is Neanderthal. Even some anthropologists wondered what our ancestors saw in Neanderthal man – was it love at first fright? Perhaps Neanderthals had novelty value because early *Homo sapiens* were black and at least some of the Neanderthals had pale skin and red hair. A few wore feathers for decoration and at least one even played a flute.

Homo sapiens were hunter-gatherers but only about ten per cent of their diet came from animals. When the hunters were not swapping stories about the mammoth that got away, they spent a lot of time stalking and losing arrow heads much as a golfer loses balls. The invention of the sling allowed the womenfolk to carry a child while harvesting nuts, grubs, roots and berries, thus ensuring a supply of essential minerals and vitamins in the diet. Because foraging leaves no distinctive artefacts some scholars have underestimated women's role. Archaeologists also love a good story and sticking a mammoth is more dramatic than picking bilberries. People were living longer and this led to the invention of the grandmother who looked after the children and cooked the dinner while mother was out foraging, which significantly increased the family's food supply.

Although the hunters couldn't out-sprint the prey, their light frame, sprung tendons and efficient sweat glands meant they were born marathon runners. They could comfortably jog behind a wounded animal until it collapsed, and still have sufficient energy to carry the joints and tall tales back home. They were nomadic and entire families followed the seasonal migrations of the herds. But with the invention of the lasso they could capture animals without harming them. Goats, sheep and cows could be hobbled or corralled close to the family shelter. Staying put meant building more permanent homes.

Our greatest invention did not arise in the cold north where nature repeatedly left the refrigerator door ajar. It happened in the near east where hunter-gatherers harvested peas and lentils and grass seeds, though they didn't know how to grow them. They cut the grass stalks with a saw-edged sickle that would not have been unfamiliar to an Edwardian farmhand. Experimental trials with ancient sickles proved that a family could harvest a year's supply of seed in three weeks. If kept dry in sealed pots, grain stored well to provide food for both humans and domesticated animals over winter. To get the edible flour out of the grass seeds they invented the quern, a stone handmill. The 'ears' were laid on a saucer-shaped rock and ground by rubbing a hand-held stone back and forth. No doubt they noticed that some seeds fell around the quern and germinated into new grass plants, but it was some time before they began to plant seeds in disturbed soil to ensure the supply. Gradually *Homo sapiens* became a settled farmer and his invention of cultivation changed the course of human history.

Domestication

The transition from nomad to settled farmer was the most important decision we ever made but at first it was not over-whelmingly beneficial. Compared with hunter-gatherers, the early farmers were smaller and showed more signs of infectious diseases, malnutrition and bad teeth. Unfortunately, the nearest dentist was over twelve thousand years away. It is probable that over time some of the wild animals they had relied upon succumbed to climate change and intensive hunting. Having domesticated animals at hand provided a surer supply of meat, milk, leather and wool, plus manure to feed the soil. A fixed home also brought social benefits for the family with both parents at home. One nocturnal benefit in particular led to a substantial rise in the population. Within a couple of thousand years farmers became dependent on cereal crops and there was no turning back.

No longer having to lug their possessions from place to place, the settlers accumulated more domestic utensils. Around eight thousand years ago the invention of the kiln enabled them to bake clay. Vessels were made by pressing soft clay over a curved object or by curling ropes of clay round and round with each ring on top of the previous one to form the walls, which were then smoothed by hand. Firing them at 600°C made serviceable but fragile earthenware jugs and cups. Early dwellings were littered with fragments of ceramics, much to the delight of archaeologists. We know what cereal crops they grew because some seeds were accidentally embedded in the soft clay.

The potter's wheel appeared six thousand years ago. It was invented by the Sumerians, who lived between the rivers Tigris and Euphrates in Mesopotamia. They were known to make 'thrown' pots, so they must have invented the potter's wheel and this may be the forerunner of all wheels. But how was the potter's wheel powered? A modern ceramicist called Bernard Leach made a wooden turntable with an upright peg on the perimeter. He used this handle to spin the pivoted wheel as fast as he could and then shaped the clay until the wheel slowed down. He spun it repeatedly until the pot was finished.

With increasing demand, specialist potters, weavers and tool makers sprang up in small factories. Some had a distinctive style and the handiwork could be attributed to a single craftsman. Wooden ploughs, adzes, mattocks and augers appeared for the first time. Large stone axe heads were smoothed and polished and fixed into long wooden hafts. I have wielded one of these formidable axes and seen what they could do. In an experiment two researchers used them to chop down young trees and made a substantial clearing in a single afternoon.

Around 5,500 years ago multiple cultures devised methods of extracting metals from their ores and stone tools were superseded by metal ones. Copper was the first to be used. It was extracted in a furnace fired with charcoal and aided by primitive bellows. The copper settled at the bottom as a 'cake'. When reheated and skimmed to remove impurities it could be poured into a mould to set in whatever shape was required. Copper 'seasoned' with tin created bronze, which is much harder than copper. The metalworkers invented 'lost wax casting'. The object, let's say a sword, was carved in beeswax and the model was encased in clay to make an impression. The mould was heated, the wax ran out through a pre-made hole and was replaced with molten bronze. When this solidified, the mould was broken open to reveal the blade, which only needed to be sharpened and polished. Swords were mass-produced by using an existing blade to make any number of moulds.

Bronze changed everything. The farmers were now self-sufficient and had surplus food to exchange for goods. It became a status symbol, and cauldrons, cups and weapons were no longer merely utilitarian but objects of beauty. Jewellery and other luxury items were in demand. 'Look what I've got,' is not a new attitude. The consumer culture was thriving long ago.

In 2007 in England a metal detectorist found almost four hundred unused bronze axes. They had probably been dumped when the demand for bronze collapsed. Around 3,200 years ago bronze was replaced by iron, one of the commonest metals on earth. It is the cheapest and also the hardest of all workable metals. An iron spear point could easily penetrate bone. Wrought iron that had been hammered into shape became the universal material for tools and weapons. In Latin the common word for a sword was *ferrum* (iron). A toolkit from this period was similar to that used by a Victorian workman.

For a long time archaeologists treated their calling as treasure hunting and the only metal of interest was gold. Then in 1816 a Scandinavian archaeologist labelled the prehistoric eras as the Stone Age, the Bronze Age and the Iron Age. It took over a century for this advance to grow into the realisation that technology and invention were at the heart of the development of human society.

The urge to build

The achievements of history are often measured in the structures they leave behind. Prehistoric societies left post holes where buildings once stood and sometimes, as at Skara Brae on the windswept coast of Mainland, the largest of Scotland's Orkney Islands, stone houses have survived. But the most puzzling structures are the elaborate arrangements of stone monoliths like Stonehenge, which seem to date from the late Stone Age to the early Bronze Age. The massive stones, weighing 2,000 tons (2,030 tonnes) in total, were hand-hauled from far away on sleds running on logs. They were raised upright by means of muscle power and planted into sockets dug with only antler picks and shoulder-blade spades.

Generations of sages have scratched their heads over the purpose of these complex structures. Even the Anglo-Saxons called them *sarsens* (troublesome stones). William Wordsworth thought Stonehenge was 'a mystery of shapes'. It has been classified as a temple, cemetery, or merely a collection of phallic fertility symbols. The latest evidence from the nearby site of Durrington Walls indicates that at midwinter and midsummer it was transiently occupied by thousands of people, some from as far away as the north of Scotland, and there were indications of great feasting. Perhaps they had invented the Rave. A gynaecologist held a different view. She claimed that from above, the site was clearly a representation of the female genitalia. I concede that a gynaecologist may have seen more female pudenda than I, but I just don't see it. Perhaps every generation gets the Stonehenge they deserve.

In 1971 I stood on a hill overlooking a sea loch at Callanish on the island of Lewis in western Scotland's Hebrides. I was surrounded by a circle of stone fangs rising from the earth. It was as if I were in the mouth of a colossal dinosaur. Thirty-seven years earlier Alexander Thom, a professor of engineering, had stood on this same spot at dusk and been so overwhelmed by the structure that he devoted the rest of his life to making systematic studies of all the 'henge' monuments he could find. In his first book he refers to no fewer than 320 sites. Thom meticulously measured the placement of each stone and concluded that even the non-circular rings were deliberate and based on smooth circular arcs or ellipses. They were executed by people who understood geometry. In addition the arrangements were based on a standard measure, multiples of what he called 'the megalithic yard' (0.829 m). This was true not just for prehistoric monuments throughout the British Isles but also those in Ireland and northern France.

Thom and others became convinced that the designers were sky gazers. Often within the circle there was a conspicuous stone that they believed was used as a sighting point from which to observe the sun at solstices and equinoxes. The ancients were certainly aware of the seasonal changes in the altitude of the sun. At New Grange in Ireland a five-thousand-year-old tomb has a passage 62 ft (19 m) long aligned so that a shaft of light illuminates the back of the tunnel only at midwinter sunrise. A beautifully corbelled structure at Maes Howe in Orkney is only a tad younger and has a similar shaft that floods the interior with sunlight on the shortest day of the year. Researchers also found numerous sites where sighting stones were aligned to conspicuous features on the visible horizon to track the movement of the sun, moon and fixed stars, far distant stars whose position appears to be stationary. In theory this would allow the observer to calculate the days in the year and even predict eclipses. They had invented astronomy.

Impressive as these monuments are, for me the most evocative reminders of ancient people are more intimate: the still visible ghost of a furrow where Wiltshire soil was turned by a Stone Age plough; the stone slab furniture at Skara Brae – a 'dresser' and a box aquarium to keep shellfish fresh – fashioned over five thousand years ago; and most touching of all, the seven-thousand-year-old burial not far from Copenhagen of a young woman thought to have died in childbirth and at her side a tiny baby cradled in a swan's wing.

THE PURSUIT OF POWER

'Forget six counties over hung with smoke.
Forget the snorting steam and the piston stroke'
– William Morris

The millennia that followed were spiced with inventors, mostly amateurs. Leonardo da Vinci's notebooks are a catalogue of ingenious ideas, but very few of his brilliant drawings were transformed into working machines because the engineers and technology of the time were not up to the job. Indeed it was not until the industrial revolution that the inventors' dreams regularly came to reality. In the following chapters I will usually begin by introducing the historical antecedents, with their successes and failures that eventually led to significant inventions.

At first all strenuous tasks demanded hard labour. The only machine available was the human body, so everything had to be lifted or hauled by hand. Even a fit, muscular man can only generate a third of a horsepower and with sustained effort his capacity and enthusiasm dwindle. The invention of the lever and the pulley gave him a mechanical advantage, but his ambition still far exceeded his ability.

Fortunately, early humans knew that there is strength in numbers. The ancient Greeks felt no need for mechanical aids, for they had squadrons of slaves to do the work. It has been estimated that it would take 100 men only twenty-five days or so to dig all the ditches and heap up the long barrow tombs at Stonehenge. Some feats required an enthusiasm for a common cause seen nowadays only in football supporters. Not far from Stonehenge is Silbury Hill, a huge man-made mound. Its construction would have taken well over a decade even if there were 500 workmen on the job.

Once animals were domesticated they lightened the burden. The power generated by a horse is, one assumes, one horse-power, although it isn't clear whether the animal in question was a giant Clydesdale or a wee Shetland pony. Animals were consumed for meat and used for all the heavier jobs, although without foresight the former might preclude the latter.

Heavy hauling was done by oxen and, later, teams of horses. To turn a millstone donkeys walked round in an ever-lasting circle or plodded to nowhere inside a treadmill. Draught animals enabled agriculture to flourish. Early farmers relied on animal dung to fertilise the soil and for house building. Two thousand years ago Lucius Columella, *the* Roman expert on agriculture, also knew what women most desired: 'He is a slothful husband who has no manure.' Over time smallholdings were subsumed into the larger estates of landowners, many of whom knew little about farming. Unfortunately, by the late seventeenth century most of England's sheep and cows *and their dung* were in the west of the country and much of the arable land was in the east, where the soil was becoming impoverished.

The soil found an unlikely saviour, Viscount Charles Townshend. He was a statesman who negotiated the peace with France and became the ambassador to The Hague, where he was a diplomatic disaster. Sir Robert Walpole, Secretary of State and Townshend's brother-in-law, took him aboard to help direct foreign policy. He soon recognised his mistake and sacked him.

Townshend then devoted himself to turnip cultivation and reinvented crop rotation. The Romans in Britain had shown that soil fertility could be sustained if the land was allowed to lie fallow for a year and then planted with beans before returning to wheat. Townshend devised a four-year cycle of crops consisting of wheat, turnips, grass and clover. The benefits of this were:

1. Changing crops prevented the build-up of pests and diseases that attacked one particular crop. The terrible potato famines that were to plague Ireland were the result of a monoculture of potatoes in the same ground year after year.
2. Wheat was a staple and its roots improved the texture of the soil.
3. Turnips were a valuable winter food for both animals and humans, though somewhat monotonous fare.
4. Most important of all, clover and the Romans' beans have the ability to convert atmospheric nitrogen into a nitrogenous fertiliser for the soil.

The viscount went down in history as 'Turnip' Townshend.

Water power

From 1760 to 1860 Britain changed from being mainly an agrarian country to an industrial powerhouse, and power supplied the driving force. Fortunately, it was a land of running water. Means of shifting water had occupied minds in the ancient world and led to several inventions, including gears. Archimedes' screw lifted water to higher levels. Even today it's used for mincing meat and 'pumping' sewage – though not both at the same time.

Waterwheels had been devised in the Middle East. The earliest model had pots lashed to the circumference of a large wooden wheel slung over a ditch. It was for scooping up water, but if it had blades that dipped into a flowing stream, the wheel turned. The horizontal axle of the vertical wheel could, by means of wooden peg gears, turn a millstone. An ancient Greek poet celebrated this advance: 'You girls who worked so hard in grinding corn – can go back to bed . . . now the nymphs of the stream leap down on the wheel to turn the axle, and with it the millstone.' Clearly he thought that a woman's place was in bed.

Over 5,600 water-driven mills were listed in the Domesday Book in 1086. Even into the Victorian era, in every village the creak, rumble and splash of mill wheels competed with bird song. But waterwheels weren't just useful for grinding cereals, they also revolutionised manufacturing.

A stretch of the Derwent Valley in Derbyshire has been designated a World Heritage Site by UNESCO. This is not for its indisputably beautiful countryside but because it embraces Cromford Mill. The vast cotton-spinning mill was built in 1771 by Richard Arkwright, an apprentice barber who became a wig maker. He made a small fortune detaching locks from the poor for pennies to deposit on the pates of the rich for pounds.

In 1769 Arkwright, who had some mechanical skill, was granted a patent on a new type of cotton-spinning machine. At that time the spinning machine of choice was James

Hargreaves's 'Spinning Jenny' (named after his dancing wife). Unlike the old spinning wheel, the Jenny could spin dozens of threads at the same time, but it was fiddly to use and demanded manual dexterity from the operator. In contrast, Arkwright's machine was so simple that it could be operated by an unskilled person. Unlike the 'Jenny', it could spin the stronger threads needed for the warp yarns through which the weft threads were woven to make textiles. The Jenny was worked by a foot treadle, whereas all Arkwright's machines would be powered by a huge waterwheel in the River Derwent.

With a superior machine and cheap labour, mostly by women and children, Arkwright could undercut his rivals. Within a few years he owned several large mills and had over five thousand employees. He added to his patents a new type of carding machine for teasing out the threads before spinning. It was originally done with the hook-covered heads of teasels.

His patents gave him sole rights to his new machine. He was willing to grant others a short-term licence at an exorbitant price – and woe betide anyone who infringed his patent. He was not a popular man in the spinning world.

As the youngest of thirteen children it was no wonder Arkwright was competitive and confrontational whenever the opportunity arose. Although he may have smiled in his youth he decided it didn't suit him and he never got the hang of humility.

His opponents united to bring him down. They challenged his patents, claiming that his inventions were not really his. On his patent application for the spinning machine he had written: 'I had by great study, and long application invented a new piece of machinery never before found out.' He also made clear that 'Richard Arkwright . . . lawfully might have and enjoy the whole profit . . . arising by reason of the said invention.'

But it was revealed in court that he had hired John Kay (the inventor of the weaver's 'flying shuttle'), who was a partner of an inventor called Thomas Highs. Kay was paid to make a surreptitious copy of all the details of a spinning machine Highs had designed. Arkwright then enlisted craftsmen to help him to build a scale model of the machine that he could use to entice investors. In addition it was suggested that the critical components of his carding machine were similar to those invented by Hargreaves. The court decided that both his patents should be rescinded.

Arkwright was angered by this verdict but it hardly affected his business and soon after he was knighted and elected High Sheriff of Derbyshire. He built himself a cosy castle to live in and when he died he was the richest manufacturer on the planet.

We don't know how much Arkwright contributed to the final design of the spinning machine, but he *was* an inventor. He devised mass production of affordable goods and invented the factory system that changed the world.

However, there was a downside to the factory system. It led to the slow decline of craftsmen and country folk were uprooted and transplanted to the treeless cities. It enriched the factory owners while perpetuating poverty in others and encouraged child labour; two thirds of Arkwright's employees were children. Profit ruled and mill owners insisted that if a spinner was ill and failed to get a substitute, he was fined six shillings a day for 'wasted steam'. Workers were fined a shilling if they were caught washing or whistling. They responded by rioting and smashing the machines that would eventually take their jobs. In France mill girls flung their clogs (*sabots*) into the works, hence 'sabotage'.

Arkwright took no chances. At Cromford he kept cannons loaded with anti-personnel grapeshot, and a posse of five thousand local men who could be mustered at an hour's notice. He had no trouble.

Yet, by the standards of the time, Arkwright was considered one of the more enlightened employers. He provided decent housing for his workers and, unlike many mill owners, he refused to employ those below six years old.

Cotton weaving became the major employment in northern England and by 1840 cotton goods made up forty per cent of Britain's exports. Cromford Mill spun its last yarn in the 1980s.

Getting up steam

The mill was driven by a waterwheel, but the industrial revolution would be powered by a different, more excitable form of water – steam. Industry was beginning to devour metals and coal and as the shallow seams ran out miners had to burrow deeper. Lower seams were prone to flooding and this could have halted industry in its tracks. It was vital to find an effective mechanism for pumping out mines. Ironically, water was the means to remove water.

As long ago as 1690 French physicist, mathematician and inventor Denis Papin demonstrated with a model machine (called a 'digester' though it digested nothing) that when a cylinder full of steam was chilled it condensed and created a vacuum that could drag a piston down into the tube. The vacuum wasn't sucking the piston down: it was being pushed by atmospheric pressure. Although we are unaware of it, the atmosphere above our head has weight and we live at a pressure of one atmosphere. This pressure would drive the early steam engines.

Ask, 'Who invented the steam engine?' and almost everyone will reply, 'James Watt.' But they would be wrong. It was Thomas Newcomen, who ran a hardware store, assisted by John Calley, a plumber and glazier, who would become the world's first mechanical engineers. Their engine was built in 1712, forty-nine years before Watt's. It was huge, with an enormous overhead beam pivoted at the centre. The rod of a piston device based on Papin's model was attached to one end of the beam, and a pump was connected to the other end. The ups and downs of the piston were transmitted to the pump by the see-sawing of the beam.

It had taken ten years of experiments and trials to perfect the device but Newcomen couldn't patent it because a Captain Thomas Savery had already taken out a patent before he had built anything. So Newcomen went into partnership with Savery to share his patent. Savery's machine was not considered to be a true steam engine because all it could do was pump water, not drive machinery. It was a slow and inefficient pump that could lift water from no more than 80 ft (24 m) down, whereas several mineshafts were five times deeper. He planned to overcome this problem by installing a series of machines at different depths in the mine. Perhaps he had forgotten that his machines were 120 ft (37 m) high and 90 ft (27 m) wide. There was no way they could be squeezed into a mineshaft.

Newcomen's machine was far more powerful than Savery's

and not as voracious, although it still consumed 6.5 tons (6.6 tonnes) of coal a day, which made it too costly to run except at coal mines. Nonetheless, for several decades it was the state-of-the-art steam machine. In 1777 seventy-five of Newcomen's engines were servicing Cornish tin mines. Six years later only one remained. They had been replaced by Watt's steam engines.

Legend has it that Watt came up with the idea for his engine when he saw the lid of a kettle being lifted by the steam from the boiling water, but it's a myth. Watt was trained as an instrument maker and had a workshop in Glasgow University. He was asked to repair a working model of Newcomen's steam engine. He not only mended it, but also decided to improve it.

Newcomen's device wasted energy by repeatedly heating up and cooling the cylinder. Watt devised a simple way of condensing the steam without cooling the cylinder at the same time, which also speeded up the stroke rate of the piston. He next increased the thrust of the down stroke by feeding steam into the cylinder above the piston. Watt patented his machine in 1769 as 'a new method of lessening the consumption of steam and fuel in fire engines'. The fuel saving was a staggering seventy-five per cent.

Watt was no businessman, so he fell in with Matthew Boulton, an engineer who could manufacture and market the engines. Within a couple of decades they sold over five hundred steam engines. In Boulton's words: 'I sell here, sir, what all the world desires – power.' Wherever there was industry there was Watt's steam engine. It was the pumping heart of the industrial revolution.

Both Watt and Boulton became rich and were elected to the Royal Society.

Watt continued to improve his engine and invented other devices, including the 'governor', which automatically controlled a machine's speed. It was an ingenious mechanical device with metal balls spinning on the ends of extendable arms which via valves could reduce the steam supply if the engine speeded up, or inject more steam if the engine slowed down. It was the first use of what we now call feedback. Watt even invented an office copier device. He also built a gas-proof chamber for Humphry Davy, the inventor of the miner's safety lamp, to experiment by inhaling laughing gas in such large amounts that he might never have laughed again. Watt coined the term 'horsepower' as a measure of power. Later the standard measure of power was named the 'watt' in his honour.

Watt relied primarily on atmospheric pressure to drive his engine and only later used steam to supplement it. He was wary of using steam at higher pressure and even dismissed a model of a high-pressure machine constructed by one of his assistants. The perpetual pessimist, he feared that in a full-scale machine the boiler would explode. However, in Boulton's factory 'Iron Mad Wilkinson' had developed hammered iron plate that was cheap and could easily confine high pressures. He loved everything about iron and built iron barges to prove that they could float. His wife buried him in an iron coffin.

The future lay in the hands of a brilliant Cornish engineer, Richard Trevithick, who designed a boiler that could withstand a pressure of 50 lb per sq in (psi) (3.5 bar), five times greater than Watt's machine. He also replaced the huge overhead beam with a neat crankshaft. His engine was therefore far smaller yet three times more powerful than Watt's. His later engines ran at an amazing pressure for the time – 145 psi

(10.2 bar). One did explode, because the operator closed down the safety valve when he went for lunch.

The advances of just a handful of inventors spurred the beginning of the industrial revolution and every kind of manufacturing was driven by steam. Only thirty years after the steam engine was adopted by industry the majority of Britain's population were factory workers. It was the workshop of the world and, with less than two per cent of the world's manufacturers, generated more than a third of the world's trade.

Small towns grew into cities. In less than eighty years Manchester, the centre of the cotton weaving industry, expanded eighteenfold. England's green and pleasant land was shrouded in acrid smoke from Blake's dark satanic mills.

In the 1840s Friedrich Engels visited Manchester and recorded his impressions: 'In a rather dark hole . . . surrounded on all four sides by tall factories . . . stand 200 cottages in which live about four thousand human beings . . . The cottages are old, dirty, and of the smallest sort, the streets uneven, fallen into ruts and without drains or pavements; masses of refuse and offal and sickening filth lie among standing pools in all directions . . .'

Some streets *were* paved with gold but only for the mill owners.

FULL STEAM AHEAD

'The great affair
is to move'
— **Robert Louis Stevenson**

As the industrial revolution got under way the demand for coal soared. The major impediment to manufacturing was transport. The roads were hardly fit to bear a coach carrying a dozen people, let alone large quantities of coal or metals.

For a while the answer seemed to be canals, for barges were far easier for horses to tow. Entrepreneurs embarked on a huge programme of canal building and hundreds of Irish 'navvies' (navigators) dug them. But canals were dogged by insufficient water during dry summers and by freezing in winter. On hilly terrain the barges had to ascend or descend in a series of locks that were costly to build. In a 15-mile (24 km) stretch of canal from Worcester to Tardebigge boats had to negotiate fifty-eight locks to rise 428 ft (130 m). Such obstacles and a reliance on horse power meant travel was very slow.

The ideal mode of transport had been known for centuries. In drift mines it was not unusual to bring out the ore or coal in wagons pushed, by a workman or even children, along wooden rails. Huntington Beaumont opened several mines in Northumberland and built 'tramways' on which pit ponies hauled coal from the pit head to coastal towns to be shipped down to London. His wagons had flanged wheels to keep them on the iron rails.

At last there was a power far greater than the horse – it was steam. James Watt had devised a means to convert the up-and-down action of a piston into the rotary motion of a wheel. By the time Watt's patent had expired Richard Trevithick was ready to fit his compact, high-pressure steam engine into a coach chassis on wheels. He was a brilliant engineer and introduced several innovations. He dispensed with Watt's heat-wasting condenser and expelled the exhaust gases through a funnel which increased the draught over the fire. His vehicle coughed smoke into the air and was dubbed the 'Puffer train', as toddlers were encouraged to call loco-motives for decades afterwards.

'Uncle Dick's Puffer' set off to demonstrate its abilities, but the tiller was temperamental. It took a liking to a ditch and the coach became the very first off-road vehicle. Within days Trevithick fixed the steering and it climbed a steep hill. At his first attempt he had produced a small engine that was powerful enough to move a vehicle. He and his mates retired to an inn to celebrate. The Puffer was parked in a nearby shed, but unfortunately they forgot to damp down the boiler's fire. The wooden hut burnt wonderfully well and the locomotive was reduced to a melted and buckled wreck.

By 1804 Trevithick had built an improved locomotive in which the flue snaked around inside the boiler so as to utilise all the heat before venting into the air. It ran 10 miles (16 km) on an iron 'plateway' (parallel strips of iron laid on wooden sleepers) and pulled five wagons carrying 10 tons (10.2 tonnes) of iron and seventy passengers at 5 mph (8 kph). It was the first-ever train ride and proof that steam locomotives could haul heavy cargoes and that there was sufficient friction between metal wheels and metal plates to make railway tracks feasible.

To publicise his machines Trevithick built a temporary, circular track in London's Euston Square on which his neat little locomotive *Catch-me-who-can* towed a modified landau carriage to give joy rides to paying passengers – another first. The advertisement ran: 'Mechanical power subduing animal speed.' It was not the great marketing success he had hoped for and perhaps this disappointment explains why he returned to manufacturing his stationary engines, which were in demand to power rolling mills in ironworks.

Trevithick was a restless, temperamental genius and once he had invented something he looked for new challenges. He was not a bespectacled boffin but a man of action, as well built as his engines thanks to weightlifting and wrestling. Precipitately he departed for Peru to survey railway routes and design mining machinery. Sadly his skills were lost to British industry.

In 1818 the Earl of Strathmore decided to build a railway from his coal mines to Darlington, the nearest manufacturing town, 37 miles (60 km) away. The wagons were destined to be horse-drawn until he met George Stephenson.

The town of Killingworth is only 4 miles (6.4 km) from where I grew up. And there a plaque on the wall of Dial Cottage records that this was where George Stephenson once lived and grew prize-winning vegetables in his garden. He was the 'enginewright' at the local coal mine and spent his evenings repairing clocks and shoes to raise the money to give his son the education that he never had. He had an intuitive understanding of machines and had built a locomotive to run on a 'tramway' from the pit head south to Newcastle.

This gruff fellow who was not good with words managed to imbue the Earl of Strathmore with his enthusiasm for the future of steam locomotives. So much so that he was appointed the engineer for the Stockton to Darlington Railway. He planned the route taking care not to impinge on the Duke of Cleveland's fox-hunting fields and set the distance between the rails at 4 ft 8½ in (1.44 m), the traditional width of space that allowed a single horse to pull a cart. It became the standard gauge for almost all subsequent railways in the British Isles.

In addition to numerous freight wagons there would also be a single passenger train. The local newspaper was not impressed: 'What person would ever think of paying to be conveyed . . . in something like a coal wagon . . . to be dragged by a steam engine?' Many people had never seen a locomotive and thought it would be an 'automatic semblance of a horse stalking on four legs'. On the inaugural run an estimated six hundred people filled the seats in the twenty-one wagons or clung to the outside, as they do in India to this day. It was the very first passenger service using a steam locomotive.

To enhance the income, the line was made available to others using their own rolling stock, most of which were hauled by horses. Unfortunately, it was a single-track railway with few sidings for passing places. To make matters worse there were no signals. The steam locomotive couldn't demonstrate its speed when hindered by all this traffic and when it had to slow down while passing through woodland in case sparks ignited the trees. Nevertheless the railway was a boon to local industry. Middlesbrough rapidly grew from a tiny village into the largest steel producer in the world.

The Stockton to Darlington Railway had been the proving ground for Stephenson's ideas; his next venture would be far more ambitious. He would be the engineer on the proposed double-track link between Liverpool and Manchester. It was an obvious development because Liverpool was the main port for Britain's transatlantic trade and two million bales of cotton left the city each year on pack horses destined for the spinning mills of Manchester.

Stephenson was alert to the power of publicity and persuaded the sponsors to offer a generous prize for the winner of a competition to select the locomotive that would be used on the line. The trials were held in 1829 at Rainhill just north of Liverpool. Five locomotives were entered but the *Cycloped* was disqualified because its power unit was a horse inside a treadmill. Stephenson's *Rocket* won hands down by towing two wagons full of rocks at 24 mph (39 kph), while its rivals broke down or exploded.

Several celebrities were guests of honour, including the Duke of Wellington. He said: 'I see no reason to suppose that these machines will force themselves into general use.' They all had a wonderful time, except for one. Some of the guests disembarked to stretch their legs. A cabinet minister called William Huskisson saw the *Rocket* steaming towards him and tried to climb back into the carriage, but fell beneath the wheels. Although another Stephenson locomotive, the *Northumbrian*, rushed him to hospital, it was in vain. Stephenson preferred to remember this not as the first rail fatality, but the day that the hospital-bound *Northumbrian* reached a speed of 36 mph (58 kph) – a new world record.

Just like flying

For a generation that knew of nothing faster than a galloping horse, speed was a concern. Stephenson assured a House of Commons committee that his trains would run at a stately 12 mph (19 kph). He lied, of course: he had no choice because 'experts' prophesied that travelling at more than 20 mph (32 kph) would suck all the air from your lungs or you would go mad. Even just watching the landscape rush by would damage your eyes. The hiss and clank of the engine would cause women to miscarry and leave the male traveller in 'a state of confusion that it is well if he recovers in a week'. Daily commuting would be out of the question.

Even innocent bystanders were in danger. A passing train could wilt vegetables in the fields, kill birds in flight and dry up a cow's udders. An objector collared Stephenson on the danger of a cow on the line with a train approaching. 'Surely,' he said, 'that would be a very awkward circumstance.' 'Aye,' Stephenson replied, 'very awkward . . . for the cow.' (Livestock and deadstock can be a problem. My wife and I took a train from Liverpool to Reading and our journey was interrupted by horses on the track, then a man's body across the line and finally a landslip that resulted in a reduction to a one-way track.)

A cartoon captioned 'The Pleasures of the Railroad' depicted an exploding locomotive with detached limbs flying in all directions from the torsos of surprised passengers. Even the guidebooks were not encouraging. One recommended that you should sit as far from the engine as possible, for if it exploded 'you would probably be smashed to smithereens'. It was best to sit with your back to the engine or you might be 'blinded by small cinders which escape through the funnel'.

It took the public some time to get used to the speed of trains. Some believe that the locomotive really did get bigger as the train approached. Others leapt from the carriage when they were close to their destination and were rewarded with a broken leg or worse. When the train reached 23 mph (37 kph) a passenger found it 'frightful . . . it is impossible to divert yourself of the notion of instant death for all'. Nevertheless, the public soon learned to sit back and enjoy the thrill of speeding 'swifter than a bird . . . when I closed my eyes, this sensation of flying was quite delightful'.

Buying a ticket was less fun. You had to book a day in advance and give your full name, address, age, place of birth, occupation and reason for travelling.

The publicity for the Liverpool to Manchester line led to a period of railway mania. Within fifteen years of the first train departing from Liverpool, Britain had 7,500 miles (12,070 km) of gleaming lines cobwebbing the country and was transporting 100 million passengers a year. Many inventors made the railway possible, but George Stevenson made it happen.

The Liverpool to Manchester line, 30 miles (48 km) long, was not easy to build. There were two long tunnels and sixty bridges to be constructed, but the biggest challenge was Chat Moss, a huge bog 'over which no human foot could tread without sinking'. Stephenson and his brilliant son Robert came up with a daring solution. They dumped into the bog the soil and rock removed when excavating cuttings. Then they made a floating raft of brushwood and heather and hoped it would support the railway embankment. And it did.

Other railways likewise faced problems arising from unco-operative topography. On the London to Birmingham route a tunnel 3 miles (5 km) long suffered from quicksand. Stephenson deployed thirteen massive pumps removing 2,000 gallons (9,090 litres) of water per minute. Even so, it took eighteen months to suck it dry. On the London to Bristol project Isambard Kingdom Brunel had to engineer a tunnel that was almost 2 miles (3.2 km) long. It took two and a half years and required 1 ton (1,016 kg) of gunpowder per week and thirty million lining bricks. No wonder Brunel smoked forty cigars a day.

Much of the construction was achieved by huge battalions of navvies with picks and spades, but by 1870 Britain's steam engines were doing the work of forty million able-bodied men – four times the entire workforce of the nation. Machines were transforming the world.

Towards the end of his distinguished career Robert Stephenson said: 'As I look back upon the stupendous under-takings accomplished in so short a time, it seems as though we had realised in our generation the fabled powers of the magician's wand. Hills have been cut down, valleys filled up . . . and if high and magnificent mountains stood in the way, tunnels of unexampled magnitude have pierced them through.'

The changes that the railway brought to society were no less than those it wrought on the landscape. The treasurer of the Liverpool and Manchester Railway summarised the trans-formation in public perceptions: 'The most striking change produced by the railway is the sudden and marvellous change in our ideas of time and space. What was quick is now slow;

what was distant is now near, and this change in ideas pervades society at large.'

The greatest effects were felt by the working class. At first the railway companies discouraged the less lucrative passengers. The third-class wagons had bench seats but no windows or roof. One company, more astute than the others, realised it was missing out on a huge number of potential passengers. So it dropped second-class coaches and upgraded the facilities on third-class to attract customers. This worked and eventually other companies did the same. When I was a lad all trains had first-class and third-class coaches and no one ever asked what happened to second class.

The railways also priced their tickets in accordance with the prosperity of the passengers, with cheaper tickets on lines serving poor areas and higher charges for the more salubrious suburbs. There were even special cheap tickets for workmen. A labourer who had never ventured beyond the boundaries of his town could now afford to travel. For a few days in the summer ordinary folk swarmed on to trains and headed for the coast. One year Brighton had 250,000 summer visitors – most of them travelling on a return ticket costing fifteen old pence. For many it would be their first holiday and their first glimpse of the sea. They were determined to have a good time and were accused of 'giving up their decorum with their rail ticket'. When the masses arrived in holiday resorts all the posh folk fled.

What time is it?

Until the coming of the inter-city railway, travellers thought of which *day* they should arrive. Now it was a matter of which hour and which minute. All of a sudden time became confusing. Until 1850 all towns in Britain lived in local time. It depended on when the sun rose, so that London in the east was more than twenty minutes behind Plymouth in the west.

This made timetabling difficult. It was already complicated because the railway companies worked on two timetables. The engine drivers used the 'operating timetable', which was significantly different from the published version. Town clocks began to display 'Local time' and 'Railway time'. Frustrated passengers were forever complaining about missing trains because they had departed early. In 1858 a passenger arrived in court too late to be heard and lost his case by default. He argued that he *was* on time by his pocket watch, which was set to the time in Carlisle, where he lived.

It was proposed that all trains should use London time, but the snag was that the time was different on either side of London. The managers of the Great Western Railway had the worst problems as they ran the east to west Bristol to London Line. Their solution was to adopt Greenwich Mean Time – except at Bath, Bristol and Chippenham! In its wisdom Parliament dismissed the proposal to convert the entire country to GMT. It took thirty-five years before the politicians changed their minds.

Speeding

When high-speed trains were about to go into service I was amused to read in the newspaper that an academic was warning that at speeds of 200 mph (320 kph) the passengers would be unable to breathe and would be asphyxiated. He didn't want to worry us even more by mentioning birds falling out of the clouds and udders failing to deliver.

Clearly he had not heard of Colonel John Stapp, who in 1954 tested seat restraints by riding in a rocket sled at 632 mph (1,017 kph) and reaching the top speed in six seconds. Yet, believe it or not, Stapp was still breathing and there wasn't a dry udder in the dairy.

LIKE A RED FLAG TO A BULL

'Glorious . . . The poetry of motion!
. . . The only way to travel . . . O Bliss!
O poop-poop! O my! O my!'
– Toad of Toad Hall
(in *The Wind in the Willows*) on seeing his first motor car

As the railways prospered, road transportation lagged behind. Surely there were horse-drawn stagecoaches hurtling down the highways. In fact, they were often slowed to walking pace because even the trunk roads were ridged by iron-hard ruts in winter and when it rained they became quagmires. Standing water often obscured the rugged surface altogether and it was not uncommon for a horse to break a leg. In winter, if the roads were passable, the passengers travelling on the roof seats cocooned in multiple layers, may have died from exposure, as indeed Keats did. There were no pavements for pedestrians. Even the ancient Minoans had a walkway, but unfortunately it was in the middle of the road.

The rocky road

The Romans built fine roads in Britain, but they had been neglected ever since. In 1767 James Watt rode on horseback from Glasgow to London because the roads were unsuitable for coaches. An effort to improve matters was instigated at the beginning of the nineteenth century thanks to Thomas Telford, a Scottish shepherd's son who became known as the 'Colossus of Roads'. By 1784 he had constructed 1,000 miles (1,610 km) of road and 1,200 bridges, including the Menai Suspension Bridge in Wales, and supervised the construction of Scotland's Caledonian Canal. His achievements doubled the average speed of horse-drawn coaches.

A successful Scottish merchant named John McAdam diagnosed the problems with British roads: they were not elevated above the level of the adjacent land, and didn't have a raised crown or a waterproof surface. He invented a surfacing method called 'macadamisation' which produced an all-weather surface. He sank his life savings into improving roads and wrote influential books on their proper construction. Eventually the government compensated him and made him Surveyor General. Much later, thanks to an accidental spillage of tar, Edgar Hooley came up with the idea of mixing tar with rock chips to make a more durable surface that he called tarmacadam, marketed as Tarmac.

Turnpikes were installed throughout Britain and stage coaches with a team of ten or more horses slashed journey times between cities. Before tarmacadam these galloping hooves dislodged the surface stones, but wheels are kinder to roads than hooves. A safe steering mechanism that became standard was designed in 1818 by Rudolf Ackerman, who sold prints for a living.

What could be more logical than to place a steam engine on a chassis to compete with the coaches? The first steam-driven vehicle to lumber along a road was invented by a

French army engineer called Nicolas-Joseph Cugnot. His 'road wagon' was built in 1770 and was clearly the result of the mating of a heavy-duty horizontal ladder and a wheelbarrow. Its enormous 400-gallon (1,820 litres) boiler was suspended in front of the leading wheel. It looked as if it was removing an unexploded bomb, which is not far from the truth because a collision would have resulted in a catastrophic explosion spraying bystanders with high-pressure steam and flying shrapnel. Cugnot's road wagon was intended to tow heavy cannons, but not very far as it had to stop frequently to take on coal, water for the boiler, and no doubt Dutch courage for the nervous driver. It progressed at the pace of a lame donkey and ran out of steam at the mere thought of hauling a cannon. Steering was limited to forcing it to go in a straight line – if it turned a corner, it fell over. A bigger 'improved' model demolished a wall. Cugnot was the first-ever person to be jailed for dangerous driving.

Curious self-propelled vehicles had been proposed in the past. In the sixteenth century Johann Hautsch of Nuremberg designed a coach powered by clockwork. It was a baroque effort encrusted with dragons and bagpipers and looked so heavy that its progress was probably measured in inches per wind of the spring. Other weird designs were boats with wheels, or powered by kites, windmill-type sails or tiny railway locomotives scurrying round a treadmill like frantic mice. The 'Push foot' was powered by steam-driven walking legs. There was also a locomotive called the 'Steam Horse' although a more apt name would have been 'Crazy Horse'. It too had iron legs at the rear that 'walked' to push it along. On its maiden outing the huge boiler exploded and the locomotive ended up lying on its back with its legs in the air. As indeed did many of the spectators.

Getting steamed up

Britain was the world centre for steam locomotion but the available engines were too complex and cumbersome to fit into land vehicles. As early as 1803 Trevithick constructed an eight-seater, three-wheeled carriage that was said to nip along at a giddy 12 mph (19 kph). Unfortunately, after several successful runs he lost interest and its engine ended up powering a metal-rolling mill.

The first commercial mobile steam engines were used in traction engines to supply a power source for agricultural machinery by means of an external belt drive. They moved from farm to farm as needed and could haul heavy loads, but they were so heavily built they had difficulty over soft earth. James Boydell tackled this by attaching five wooden boards to each wheel so that the vehicle effectively laid its own road as it progressed. It was marketed as having 'endless wheels'. The multi-purpose 'Darby Digger' had no wheels, only huge downward-pointing forks on which it 'walked' as it dug. This required a wondrous array of crankshafts and eccentrics, the second being devices that convert rotary motion into linear motion. Excursions on to the road were discouraged because it ruined the surface.

Meanwhile Walter Hancock's steam-driven omnibuses ran scheduled services in London and beyond. His buses were given charismatic names such as *Infant* and *Autopsy*. They touched 20 mph (32 kph) on the best roads but all steam-driven vehicles were accident-prone. Their noisy approach spooked horses, causing them to run amok. One 'steamer' lost a wheel and collapsed on its boiler, which then exploded, dispatching all its passengers to their final destination.

Such incidents aroused an anti-steamer lobby. James Watt was so scared of steamers that he banned them from approaching his house. Many people thought steamers were noisy, dirty and unreliable – whereas horses were just unreliable and dirty. London's streets were clogged with many tens of thousands of horse-drawn vehicles and the horses inevitably answered the call of nature while working. Colossal quantities of horse dung meant the main shopping streets had to be cleared every night to make them passable by morning.

In fact, horse-drawn traffic was seven times more dangerous than motor vehicles. Sir Joseph Banks, the famous botanist and explorer, had a huge carriage. A cautious man (except when cohabiting with Tahitian maidens), he had all the latest safety devices. Should the horses bolt, 'There was a drag chain . . . to obviate the possibility of danger going down hill – it snapped, however, on our first descent; whereby the carriage ran over the post boy . . .'

Applying the brakes

By 1840 Britain had 22,000 miles (35,400 km) of turnpike roads with eight thousand toll gates charging prohibitive rates for self-propelled vehicles – at least ten times higher than for horse-drawn carriages. Saboteurs dumped thick layers of stones on to roads to halt the steamers, but they ploughed straight through while the horse-drawn carriages struggled.

Lobbying of Parliament led to the 1865 Locomotives on

Highways Act. This dictated that self-propelled vehicles must have a crew of three – a driver, a boiler man and a third man to walk ahead of the vehicle with a red flag to warn of its approach. The last hardly seemed necessary as the vehicle was not allowed to exceed 4 mph (6.5 kph) on country roads and 2 mph (3.2 kph) in town. This uniquely British law virtually killed off any research into road vehicles and didn't prevent accidents. While crossing the road Bridget Driscoll failed to see a car dawdling towards her. She became the first pedestrian in Britain to be killed by a horseless carriage. The coroner hoped that 'such a thing would never happen again'. Automobiles went on to kill more people than all the wars in the twentieth century. The 'Red Flag' act was not repealed until 1896, by which time the motor industry was well established in continental Europe.

Internal combustion, external explosions

In 1860 Étienne Lenoir, a Belgian inventor, made a more compact steam engine that sucked air and coal gas into a cylinder, where it was ignited by a spark. The ensuing explosion moved the piston. Thank goodness he didn't take up a suggestion from Christiaan Huygens (inventor of the pendulum clock) to use gunpowder as fuel. Lenoir had invented the internal-combustion engine, but it was inefficient and liable to overheat and seize up. It also took over three hours to travel just 6 miles (10 km). Lenoir realised that being dependent on gas from the mains supply might limit the car's touring range. It was left for others to exploit his ideas.

Nikolas Otto was a German school dropout who worked

in a grocery shop and became a travelling salesman and a self-taught mechanic. He saw Lenoir's engine at an exhibition and felt sure he could improve it. Having built a partially successful engine nicknamed 'the rattling monster', he and his partner Eugen Langen built the first practical four-stroke internal-combustion engine to run on gas. Although it was a stationary machine it set the template for car engines of the future. It was more compact than Lenoir's engine and consumed half as much fuel. Its fuel economy won the Gold Prize at the Paris International Exhibition of 1867. Later improvements included cooling the cylinders with water jackets.

Within a decade Otto sold over thirty thousand engines for use in factories. His production manager was a young ex-gunsmith named Gottlieb Daimler. An ambitious fellow, Daimler and his chief mechanic, Wilhelm Maybach, set up on their own to design petrol-driven engines. He replaced Otto's permanent flame ignition (like that on a gas stove) with electrical ignition of the petrol vapour, for which purpose he invented a carburettor to atomise the petrol and regulate the air/petrol mix. To dispel the public's concern over engines that relied on explosions, Daimler wound wire around the door handles so that it would look as if it were powered by electricity. The jovial workaholic worked all night behind locked doors. He was so covert that the gardener reported him to the police, thinking he was a counterfeiter.

The prototype engine had a much faster action and more power than Otto's. Daimler fitted it into a two-wheeled frame and his sixteen-year-old son Paul became the owner of the very first petrol-driven motorbike. But he was in no danger of becoming a 'Wild One' – not with wooden wheels and stabilisers.

Daimler's aim was to sell an all-purpose engine that could be fitted into whatever form of transport the customer desired. As a demonstration piece he bought a Phaeton, an open two-seater carriage minus the shafts and horses, and fitted his engine and transmission into it. It was the epitome of a horseless carriage; it still had the whip-holding brackets. Daimler had invented the engine of the future, but not the vehicle to carry it. That would come from another German inventor, Karl Benz.

Benz was also a former mechanical engineer in Otto's factory who had struck out on his own. His *Motorwagen* was nothing like a carriage, nor did it resemble a car. It was a lightweight three-wheeler with spoked bicycle wheels. It looked like a fat man's pram but it had been designed specifically to take the machinery. Benz included his bevel gear differential, which compensated for the different distances the outer and inner wheels had to travel when turning a corner. The Motorwagen was marketed as 'an agreeable vehicle as well as a mountain-climbing apparatus'.

At first its reliability was suspect. A fellow motorist wrote: 'Whenever we met a motorist hung up by the road side . . . be sure it was a Benz in difficulties with its ignition . . . The proprietor invariably wore the air of a man who was looking for a mutton chop he had mislaid – and would take some three weeks to find.' (Even more modern cars could be plagued with such problems. Long ago my fiancée was driving me deep into the countryside at night for who knows what nefarious purpose. The rain was hammering on to the roof like bullets when suddenly her early Mini stopped and all the lights went out. 'I think I saw a garage a mile or so back,' she said. Without a raincoat I got out of the car into the storm and slammed the door behind me. Instantly the lights came on and I swear the engine chuckled as it was restarted.)

Benz didn't sell a car for two long years. When he resolved the ignition problem and marketed the four-wheeled Viktoria, his name became a watchword for reliability. In 1958 one of his cars from the 1890s that was preserved in London's Science Museum was still fit enough to take part in the London to Brighton rally and completed the trip at an average speed of 8.5 mph (13.7 kph).

Benz's cars were underpowered, whereas the Daimler Motor Company was making ever more powerful engines. Their 3.5-horsepower motor ran twice as fast as Benz's and it was reliable. Over a hundred competitors started the first race for horseless carriages, from Paris to Rouen. Every one of the vehicles that finished had a Daimler engine. Daimler's other inventions included the clutch and gears that would give different forward speeds, not bad for someone Otto had called 'indescribably thick-headed'.

Reliability was key to selling cars. In 1895 a car trial in Chicago was watched by Hiram Maxim (the inventor of the machine gun). He described the vehicles as 'an astounding assortment of mechanical monstrosities . . . every machine needed about five hours of tinkering for every hour of running'. Only two reached the finishing line. One was a vehicle made by the Duryea brothers which became the first successful American automobile. In 1908 a French newspaper sponsored a marathon rally to show the reliability of French cars. Unfortunately, no French cars finished and the German team that came in first were disqualified for hitching a lift on a train.

Most of the mechanisms of future motor cars had been invented, some of them long ago. In 1675 Robert Hooke devised the propeller shaft with universal joints that allowed the axle to rise and fall on a spring. Even stage coaches had primitive brakes, headlights and horns, adjustable windows and sprung suspension.

In 1891 two French tool makers, René Panhard and Émile Lavassor, built a motorised dog cart. It was a flop (perhaps it hesitated at every lamppost) but their next effort *looked* like a motor car, for the engine was up front under a box-like bonnet. They also replaced the tiller found in most horseless carriages with a steering wheel. Their compatriot P. de St Senoch devised a hydraulic suspension system to replace springs in 1906. Fifty years later the French car manufacturer Citroën revived the idea. Similar hydraulics were also responsible for the wildly bucking limos beloved by LA pimps. Triplex safety glass came out in 1909. It was composed of a thin sheet of celluloid sandwiched between two sheets of glass. Unfortunately, the celluloid tended to craze and turn yellow, so some drivers preferred to put their head out of the window to get a clear view. Non-yellowing safety glass wasn't available until 1936.

Luxury at last

All of these developments were brought together in the first truly modern automobile, a Daimler that appeared in 1901. Emil Jellinek, Austrian consul at Nice, had a concession to sell the new model in France. He had lots of wealthy friends and relations, so he acquired the first thirty cars and the new marque was named after his daughter Mercedes. It was an immaculate beast with a 5.9-litre, four-cylinder, thirty-five-horsepower engine and a gear stick that slid through gates so that any gear could be selected without going through all the lower gears. At the nose of its long bonnet was a radiator to cool the engine. The Mercedes was sold as 'the car of the day after tomorrow' and became the car of choice for the British royal family.

In 1901 Britain had over three million horses and only 304 car owners. But that was about to change. However, which power source would propel the automobiles of the future was still uncertain. It was a three-horseless race between steam, internal combustion and, believe it or not, electricity. Electric cars were pioneered in the 1890s by Lenoir. According to a contemporary commentator they gave the 'most freedom from care of all motive power'. They were simple to drive, had 'few parts to cause apprehension, and any derangement was soon remedied with but a small practical knowledge of the wiring connections'. Their main drawback was that the batteries were heavy yet could not store enough power for long journeys, and there were too few charging places. Nevertheless, electric runabouts and taxis became popular. A motoring book of 1901 devoted a hundred pages to electric vehicles and illustrated forty models. It even had a chapter on 'How to build an electric car'.

In 1903 a Belgian racing driver exceeded 62 mph (100 kph) in an electric car. But eventually there were only electric trams and trolleys fed from overhead wires. Although when I was a boy all the milk floats were battery-driven because their stop/start silent running didn't wake anyone on their early-morning rounds.

In 1902 just over nine hundred motor cars were registered in New York State, but more than half were steamers. The invention of the 'flash boiler' had solved the problem of the need to wait for the engine to get up steam. At the beginning of the twentieth century the boiler's inventor, Léon Serpollet, drove a steam-powered 'Sprint Racer' at 81 mph (130 kph). In 1906 a Stanley steamer raised the record to 127 mph (204 kph). As steam vehicles became more numerous, pedestrians and motorists were enveloped in smoke, the latter by the output from the chimney of the steamer in front.

Surely the internal-combustion motors were superior; they invariably did best in trials. Yet there were objections: 'Those who look upon them as the ideal source of energy for motor carriages will find themselves greatly mistaken.' Apparently they occasionally 'refused to start for no reason', were dependent on gear changing and discharged 'evil smelling and poisonous vapour'. Nevertheless, the petrol-driven automobile triumphed.

The motor car released the traveller from the restrictions of the timetable. He now could roam the highways at will and spend his weekends grinding gears and cranking the starting handle. A Jeremy Clarkson of the day wrote: 'Being drawn along without horses is a pleasurable sensation which grows until the driver is carried away with it.'

Although a driving test was far away in the future, there was a need to drive carefully until 'the horses were reconciled to the new state of affairs'. A handbook gave hints such as knowing the extent to which the carburettor, which controlled the ratio of air to petrol, needed to be 'tickled' to ensure starting: 'Yank the starting handle upwards and if the engine does not start, have another tickle.' There are also twelve pages on how to change gear, which was clearly much trickier before synchromesh gears were invented. If a horse was crossing in front of you and you were going fast, you were advised to 'go for the opening and try to dash through'.

Accidents were commonplace. In 1885 Benz was so intent on waving to onlookers that he forgot to steer and lost an argument with a brick wall. Puccini the Italian composer and a motoring enthusiast, almost failed to finish writing *Madama Butterfly* when he was involved in one of the first traffic accidents in Italy.

A motoring book assured the driver that the 'greatest risks entailed in motoring is the danger of catching a cold' because of the high wind speed and rain. The best defence lay in the correct attire: 'The coat should be long and full at the bottom', but 'beware of its tendency to distend like a balloon'. Leather garments were 'neither essential nor healthy because they confine the body's exhalations'. Similarly, 'wearing a mackintosh is most unhygienic'. The author recommends an apron to wear over the lap against the rain and, more mysteriously, 'so no wet can penetrate from below'.

Anti-dust masks transformed the driver into the most sinister alien *Doctor Who* ever confronted. Ladies were recommended to adopt a dustproof veil in the style of the face stocking worn by a bank robber. Sadly, if she went motoring regularly she 'must relinquish all hope of keeping her soft peach-like bloom'. The remedy was 'a cold rough towel . . . not used sparingly'.

A large four-seater with an annual mileage of 4,000 miles (6,440 km) cost only £11 a year to run, and three pence for petrol. The most expensive item was the chauffeur at £52 a year – still, it was £13 less than a skilled coachman used to get. Clearly motoring was a rich man's pastime. Limousines were often custom-made for their owner. One gentleman specified a silver wash basin with running hot and cold water, a wine cabinet, and sumptuous armchairs that converted into a bed in case he was accompanied by his secretary. A more down-to-earth customer insisted on a flush lavatory. My favourite was an Indian potentate who had the entire bonnet shaped like an enormous swan with eyes that lit up at night. The exhaust gases blew a whistle in its beak. The first four-wheel-drive vehicle could pass through an ordinary doorway and climb stairs, so it could be parked in your upstairs apartment.

Henry Ford

For the man in the street owning a car was just a dream – until Henry Ford came along. He was an American farmer's son who hated farming, but liked tinkering with machinery. He was apprenticed to a machine shop and rose to be chief engineer at Detroit Edison, an electricity company.

When he was forty Ford broke free to make racing cars. This financed his real ambition – to build a car sturdy enough to last for four years and so cheap that the owner could afford to buy another. His first car had a bicycle seat, no brakes and was too wide to get out through the workshop door. In 1908 he marketed the Model T, nicknamed the 'Tin Lizzie' although it was made of vanadium-alloy steel and built to withstand rough roads and inept drivers.

Ford realised that the way to make cheaper cars was to

cut production costs. He did this by buying in standard parts from other firms and assembling them in his factory. But instead of bringing the fitters to the car, he brought the car to the fitters.

He didn't invent either mass production or the assembly line, but he refined them. Over a hundred years earlier Eli Whitney (the inventor of the cotton gin) received an urgent order for ten thousand muskets. Instead of each worker making a gun, he made identical replicas of a single component, which took far less time. Whitney demonstrated how ten muskets could be rapidly assembled by picking parts at random from a pile of parts.

Ford knew that assembly lines operated at Sears Roebuck's parcelling factory, and in the Chicago slaughter houses, where it was more of a disassembly line. The Model T was said to have about five thousand parts and each one had to be fixed on to the car. Each task, no matter how small, was done by a different person. There was a bolt pusher-in, a nut-placer, a nut-tightener and so on. All this was done with the car moving on a conveyer belt, the speed of which was controlled to allow particular tasks to be done, but without a second to spare. The well-paid fitter needn't be highly skilled, only inured to monotony.

A new car was driven off the conveyer belt every ten seconds: up to two million cars a year. Ford revolutionised car manufacturing and created a mass market for private cars. An entire generation of optimistic men learned to canoodle and face disappointment in the Model T. The world became a paradise scented with the fumes of petrol. As the internal-combustion engine raced into the future, the steam car became a distant memory.

Well, not quite. In occupied France during the Second World War the eighty-year-old Marquis de Dion, an automotive pioneer, resurrected his 1889 three-wheeler and steamed around petrol-starved Paris annoying as many Nazis as he could find.

RUBBER DUB DUB

'On a round slippery wheel that rolleth,
and turns all states with her imperious sway'

– Sir John Davies, circa 1600

The earliest evidence of the wheel is a five-thousand-year-old Sumerian ceramic in the shape of a wheeled cart. While celebrating the unknown inventor of the wheel, we should also offer belated congratulations to the person who added the other three. The first clunky cart wheels were just several boards of wood held together with cross batons. Improvements involved steaming a strip of wood and fixing it to the edge of the wheel with copper nails to make a rim. Later metal rims became the norm and radiating wooden spokes gave more spring to the wheels. Lightweight wheels with thin spokes were a feature of the agile war chariots of the ancient Egyptians.

Bicycle days

Early bicycles had wooden wheels and wooden frames too. The first models had no pedals: the rider simply sat astride and walked, like modern 'balance bikes' for toddlers. It was frighteningly faster than walking pace when going downhill, but far, far slower going uphill. Worse still, one inventor forgot to provide any means of steering the bike.

Bicycles became known as 'bone shakers'. Sore-bottomed cyclists demanded something to absorb the bumps in the road surface and it came in the shape of better saddles, thin wire spokes and solid rubber tyres.

It is amazing that the manufacturers persuaded people to attempt to balance on a dangerously unsteady machine with only two wheels. Yet the most popular bike of its time was the most dangerous of all. Before chains and gears were invented the speed at which a bicycle progressed for each turn of the pedals was determined by the diameter of the front wheel, so the obvious thing to do was to make that wheel bigger. James Starley, who worked in a sewing machine factory, came up with the Ordinary bicycle, which could not have been more extraordinary. It became known as the 'Penny Farthing' because the front wheel was 5 ft (1.5 m) or more in diameter while the rear, trailing wheel was tiny. The cyclist-cum-high-jump champion had to leap on to the lofty saddle while getting the bike to start moving. From his perch he had a wonderful view of the passing landscape, even surreptitious peeps into bedrooms. Not infrequently he also got a sudden close-up of the road surface. Applying the brake while going downhill was not to be recommended. In 1884 a marathon cyclist named Thomas Stevens rode a Penny Farthing across the USA. It was not just the tyres that got blisters. Penny Farthing riders were not popular with other road users. One who had the audacity to overtake a coach was wrenched from his bike and soundly whipped. One driver's sport was to throw sticks into the spokes and watch the rider fall off.

James Starley's nephew produced the Rover safety bike – safe because the rider could put both feet on the ground. It was the first modern-looking bicycle to have same-sized wheels with overlapping tangential spokes – that is, they were not fitted radially but at an angle to the hub to make for a much stronger, well-sprung wheel. There was also a chain drive connecting the pedals via a toothed chainwheel to the rear hub. Three hundred years earlier Leonardo da Vinci had drawn a similar articulated chain.

Cycling became the most popular form of transport and *the* great weekend leisure activity. In 1892 it was estimated that there were over a thousand Sunday cyclists on the London to Brighton road. Among the hordes of men there were some emancipated women who cycled. They were not, however, so emancipated that they would cock their leg over the crossbar, so a ladies' model was produced. As women rode side-saddle on horses, so should they sit sideways on a bike – with only one pedal. It would need a contortionist to ride such a ridiculous machine. Women adopted a more practical solution for cycling without any loss of modesty. They wore baggy pantaloons which were popularised by Amelia Bloomer, hence 'bloomers'.

Tyresome

In those days tyres were made of rubber from tropical trees that bled sticky latex when the bark was cut. The Aztecs made bouncy balls for games out of it. Explorers brought samples back to Britain and the famous scientist Joseph Priestley found that when it was rubbed on pencil marks these were erased, so he named it 'rubber'. Not everyone in the world uses this word when referring to an eraser, as I found out in the United States when I asked a secretary if I could borrow a rubber. When she blushed and demurred I reassured her with: 'I only want it for a few minutes. I'll give it back to you when I'm finished.'

Unfortunately, if rubber got warm it went soft and sticky and when cold it became brittle. Charles Goodyear, although a bankrupt with no chemical expertise, determined to stabilise rubber and make his fortune. After he'd tried mixing rubber with everything from castor oil to cream cheese, serendipity came to his rescue. He accidentally spilt a mixture of rubber and sulphur on to a hot stove, and when scraping it off he noticed that the edges were elastic and not tacky. Further experiments revealed exactly how to produce this new tough, pliable material, which he called Vulcanite.

He borrowed $30,000 to fund a pavilion in the Great Exhibition of 1851, held at the Crystal Palace in London. Ingeniously, everything inside his pavilion was made of rubber, but perhaps it would have been better to have highlighted its *useful* applications. There would never be a big market for rubber musical instruments.

Goodyear patented his method, but several people guessed his recipe and went into production, and he exhausted his profits fighting patent infringements. For his achievements he was awarded the Légion d'Honneur, which was sent to the debtor's prison where he was lodging yet again. It was said that: 'If you meet a man wearing an India rubber cap, coat, vest and shoes, with a rubber money purse without a cent in it, that is he.'

Yet he didn't anticipate the most lucrative use of all – tyres.

The massive Goodyear Tyre Company is named after Charles although neither he nor his family were ever connected to the firm.

Solid rubber tyres were fine, but if they were hard enough to take the wear they gave a relatively rough ride. In 1840 Robert Thompson, a Scottish engineer, came up with a better idea. He patented a design for an air-filled tyre for the wheels of horse-drawn carriages, but his 'aerial' tyres kept falling off the rims.

Forty-eight years later another Scot, veterinary surgeon John Dunlop, was dismayed to see the solid rubber tyres of his son's tricycle gouging the lawn. He thought of replacing them with a sealed hosepipe full of water. A friend suggested air would make a better cushion and Dunlop later patented an inflatable inner tube protected by an outer rubber casing with a tread that promised a smooth ride with 'Vibration impossible'. The inflatable tyre was eagerly embraced by cyclists, who in future would halt not just to admire the scenery, but to mend a puncture.

The nascent car industry also adopted pneumatic tyres. Dunlop benefited not just from direct sales of tyres. On the island where I live his franchise used to supply all the tyres for Isle of Man Transport and charged one and a half pence per mile. Dunlop made a fortune and the company he founded is still going strong. But it's not the most prolific manufacturer of rubber tyres. A Danish company makes 306 million little tyres a day. The company is LEGO.

The demand for tyres became huge and almost all the rubber for the American automobile industry came from abroad. Sometimes the supply was erratic, so Henry Ford had a radical idea. He would grow his own.

It was not his only venture into horticulture. He grew three hundred varieties of soya bean. To persuade doubters of the vegetable's virtues he held a feast with soya in every course from the soup to the dessert. At the event he wore a trendy soya suit and tie.

Fordlandia

In 1928 the Ford Company bought a concession for an area of the Amazonian forest larger than many American states. It included a tributary of the River Amazon so that it could be serviced by the company's fleet of ships.

When I was a boy I read a book called *Exploration Fawcett*. It was the true story of Colonel Fawcett, who entered the forest not far from Ford's plot to seek a lost city. He was swallowed by the forest and never returned.

The Amazonian forest was a lively place inhabited by ruthless and armed plantation owners, and natives who were handy with curare-tipped darts, known as 'flying death'. Wildlife was plentiful. The rivers were crowded with grinning piranhas. The anacondas were big enough to strangle an ox and swallow it whole. And at night vampire bats went out on the town. Now there was a new town to haunt. Welcome to Fordlandia.

Ford was so rich he could simply sweep away the forest and plant rubber trees in their hundreds of thousands. Despite its isolation Fordlandia had almost everything that you would find in a typical mid-western town. The houses had electric light, refrigerators and flush toilets. Model Ts tootled down the streets and at night street lamps came on. The residents could visit the local picture palace to see a movie and at weekends they would enjoy a day at the swimming pool or play a round on the golf course. There was even a hospital.

Ford's multi-million-dollar project was not just a business venture. His Shangri-La outpost, based on old-fashioned virtues, honest labour and Utopian capitalism, would show America what it *should* be like. It would be a jewel in the jungle. Sadly, it was a disaster.

The close planting of rubber trees allowed leaf blight to destroy the crop. An entirely new plantation and a second town had to be built. The local workforce was unruly and opportunists opened speakeasies and brothels. The management struggled to keep order. In addition many workers were bitten by poisonous snakes and by the numerous mosquitoes that carried malaria and yellow fever. Fordlandia was abandoned in the late 1930s and now its only inhabitants are bats. The shortage of rubber during the Second World War stimulated the development of synthetic rubber, which is what modern tyres are made of.

Ford was a vehement pacifist who turned to manufacturing tanks when the United States entered the war. He swore that he wouldn't take a cent in profit. True to his word he didn't make a cent; indeed it was estimated that he made $29 million.

Ford was a man of strong opinions. When addressing patients with heart problems he told them to throw away their medication and start chomping celery instead. He was also convinced that all criminals started out as smokers and apparently Jews were responsible for all things evil, including jazz and short skirts – two of my favourite things.

MESSING ABOUT IN BOATS

'Speed, bonnie boat, like a bird on the wing'
– Harold Boulton, 'The Skye Boat Song'

We must now reach back into history long before the first rubber dinghy was inflated. Early humans were great walkers but they weren't entirely land-bound. Indeed some of the ancient migrations required them to cross water. A 10,500-year-old paddle was found in Yorkshire and a dugout canoe only a thousand years younger was discovered in the Netherlands. Boats can be made from a variety of materials and early migrants did not confine their activities to lakes and placid streams.

Drifters

In 1948 Norwegian anthropologist Thor Heyerdahl, intrigued by similarities in ancient populations separated by oceans, crossed the Pacific Ocean on a raft made from balsa-wood logs to show that ancient peoples could have made such a crossing. Later he traversed the Atlantic from Egypt to South America on a tiny boat made entirely of bundles of papyrus reeds. A Briton, Tim Severin, followed in Heyerdahl's wake by sailing an Irish *curragh*, an open boat with ox hides sewn on to a wooden frame held together with leather thongs, across the Atlantic to Iceland and then on to Newfoundland. More recently he built a raft entirely from long bamboo poles lashed together with rattan. It departed from Hong Kong and drifted 5,500 miles (8,850 km) across the Pacific. All these craft proved safer than the *Costa Concordia* cruise liner.

Both Heyerdahl and Severin were testing their theories on human migration and they proved that despite storms and deprivation these tiny, frail craft could carry people across vast oceans. Some anthropologists are convinced that early man in South-East Asia and Australia made long-distance seafaring migrations.

Those early primitive craft were driven by sea currents and the wind in their sail. A single square sail to harness the wind was for centuries the norm on seagoing vessels. But the wind is fickle and such a sail could only push the boat downwind. This was less than ideal if the mariner was aiming for somewhere in particular. Every yachtsman knows that tacking – that is, following a zigzag course – steals some forward motion when the wind is blowing sideways to the boat, but this ploy does not work well with a square-rigged sail.

Fortunately, manpower and the oar were available. The body of the pharaoh Cheops was rowed up the Nile on a funeral barge to his waiting tomb. The disassembled barge was found by archaeologists and is now restored and on display in a museum beside the pyramids at Giza.

The Chinese shunned the oar in favour of pedal power. They built huge river warships, some with a crew of eight hundred men, two hundred of whom trudged around inside treadmills to turn twenty or more paddlewheels on each side of the ship. These 'flying tigers' were said to run 'like the wind'.

The ancient Greeks mustered the greatest oar-power ever seen. The trireme was a mighty battleship that rammed enemy boats to sink them. It was driven by three banks of oarsmen, one above the other, with eighty-five rowers on each side. They powered along at almost 9 mph (14 kph) for sixteen hours at a time – a superhuman feat that modern athletes couldn't match.

The Vikings' open longship had a small square sail which carried them to Iceland and Newfoundland five hundred years before Columbus even got out of bed. But the longship's main propulsion came from the oars. It was steered by changing the stroke rate on one side of the boat and with a single large oar held upright in the sea near the stern. This oar led to the term starboard (a corruption of 'steerboard') for the right side of the boat as you face the bow. The longship could brave the wild ocean and sneak up rivers for a surprise attack.

By the thirteenth century most ships had a keel, a tiller-operated rudder and one or more square-rigged sails on a mast amidships. However, the triangular, curving lateen sail familiar from Arab dhows could sail much closer to the wind than a square sheet, so shipwrights added triangular sails – at first one, later several – to the rigging. As ships got progressively larger they began to carry an immense area of canvas and a large crew was needed to tend the rigging. Sailing ships ruled the seas for centuries.

Pioneer steamers

When Napoleon was told of steam-driven boats his response was: 'Would you make a ship sail against the winds and currents by lighting a bonfire under her deck? I have no time to listen to such nonsense.' Steam engines were heavy and bulky and had a hearty appetite for coal, leaving less space for cargo and passengers. Even so, in 1787 John Fitch, an American, installed a 7-ton (7.1 tonnes) steam engine into a narrow hull to power mechanical paddles along both sides of the vessel. His design was clearly inspired by native canoes, but it looked like a robot centipede.

After this fiasco Fitch naturally turned to a local clock maker for help in building a modified waterwheel to go on the stern of his next boat. In 1790 he launched the first ferry service, linking Philadelphia with Trenton, New Jersey, but the boat suffered from mechanical problems and the sailings were infrequent, so the venture failed.

The spotlight turned to Scotland, which would soon become the shipbuilding capital of the world. The stimulus for the first steamboats came from Patrick Miller, a banker and landowner who had ideas and the means to realise them. He came up with the ingenious notion of a twin-hulled boat (a catamaran) powered by paddlewheels in the gap between the hulls. Initially he enlisted thirty men to turn a capstan to drive the paddles. Among their number was the poet Robert Burns, one of his tenants. Manpower failed to do the job, so Miller hired an engineer, William Symington, to replace the sweaty men with a huff-and-puff steam engine.

The trial was satisfactory but there were teething problems with the engine. Symington wrote to James Watt for advice and received a tetchy reply advising him that his engine infringed Watt's patent and unless he ceased immediately he and Miller would find themselves in court. At this point Miller's enthusiasm for power boating drained away and Symington was sacked.

Even without a sponsor Symington continued to improve his engine and patented a simple and efficient system of linking a steam engine to a paddlewheel that would become standard. This attracted an influential backer, Lord Dundas Melville, the First Lord of the Admiralty and also a director of the Forth and Clyde Canal Company. He commissioned Symington to design a tug boat to replace the horses that hauled barges through the canal. Symington produced a sleek steamer 56 ft (17 m) long, driven by a twelve-horsepower engine. It was named *Charlotte Dundas*, after his Lordship's daughter. On its first working day it towed two fully laden barges almost 20 miles (32 km) against a strong headwind.

Symington received an order from the Bridgewater Canal Company for eight similar tugboats. He was on the brink of a successful career, but fate stepped in to crush him with her hobnail boots. The Duke of Bridgewater died suddenly and the order was cancelled. No other orders came and Symington died 'in want'.

The *Charlotte Dundas*, today celebrated as the first reliable and successful working steamship, was decommissioned because its wash was eroding the canal's banks. It was abandoned to rot in a creek.

Henry Bell, another Scottish pioneer of steam boats, was full of energy and ideas. In 1800 his proposal to the Admiralty for a steam-driven ship was rejected. Three years later he tried again but even though Admiral Lord Nelson was supportive the Admiralty could see no future for steam-powered ships. Nelson would become blind in one eye, but the Admiralty was blind in both. Bell designed and built the

Comet. There were sealing problems with the numerous flues inside the boiler, which were solved by using 'a liberal supply of horse dung', a common sealant in those days. The tall chimney doubled as a mast for an auxiliary sail.

Cruising

In 1812 Bell began a regular, timetabled service of three departures a week from Glasgow to Helensburgh and Greenock via the River Clyde estuary. He had inaugurated what would become the Glaswegians' favourite excursion 'doon the watter'. The smooth boat cruise was two hours shorter than the ordeal of jolts and 'shuggle' of the rocky road journey. The *Comet* offered a salon with 'periodicals and interesting books' and the private cabins had beds – for a four-hour journey? Was he expecting honeymooners?

After seven successful years on the Clyde the *Comet* was given the job of linking Glasgow to the Atlantic-facing west coast of Scotland, the first scheduled seagoing service in the world. Brunel, the great railway and ship builder, said: 'Bell did what we engineers failed in. He gave us the sea steamer; his scheming was Britain's steaming.' The *Comet*'s engine is still to be seen in the Science Museum in London.

While visiting Europe in 1805 an American named Robert Fulton examined the recently laid-up *Charlotte Dundas* and had its workings 'explained to him by Mr Symington himself'. They also corresponded and what Fulton gleaned was put to good use when he returned home the following year.

He thought of himself as a mechanic, so much so that he described his newborn son thus: 'Every wheel, pinion, screw, bolt, lever and pin about him is in the best proportion, size and strength. This has been a successful experiment' – if not a very cuddly one.

Fulton has been called the inventor of the steamship, but he added nothing to the engineering. His ships were powered by Boulton and Watt steam engines bought off the shelf and the link to the paddlewheel was Symington's design. What he *did* do was to launch steam-powered water transport in North America.

He began his working life as a silversmith and painter of miniature portraits. Indeed his trip to Europe was to improve his art. He 'spent most his time feeling sorry for himself and seeking patrons' until he met Robert Livingston, American consul in France, who had control of transport on the waterways of New York State. He contracted Fulton to design steam-driven ferries for the River Hudson. When under way his boats vomited sparks and smoke and were likened to 'the Devil going up the river in a saw mill'. Even so, they were an immediate success.

When returning from a hearing relating to a charge of patent infringement, Fulton's party crossed a frozen lake and his lawyer fell through the ice. Some might have happily left a lawyer to drown, but Fulton hauled him out, caught pneumonia and died.

Sailing ships slowly gave way to steamships. At first steam engines were fitted to fully rigged sailing ships, although sparks thrown on to the canvas was not an ideal arrangement. Such 'hybrid' ships were still wind-driven and used their engines only when in harbour. The *Savannah* was falsely credited as the first steam ship to cross the Atlantic. To voyage to England the captain filled all the passenger accommodation with wood and coal. Even so, the fuel was used up by lunchtime on the fourth day, with twenty-three days still to go.

Brunel's prize

Crossing the Atlantic on steam was a coveted prize and Isambard Kingdom Brunel was keen to carry it off. He saw Bristol, the terminus of his Great Western Railway, as the gateway to America. So he set up a shipping company and designed and built the *Great Western*, a steamship big enough to carry enough coal to cross the Atlantic. It was a 'hybrid' but the sails were just for show, for it would be propelled by enormous paddlewheels and driven by the largest steam engine ever built.

A rival firm chartered the *Sirius*, a ship half the size of the *Great Western* and loaded to the gunnels with coal. The *Sirius* had a four-day start and only just made it. It had to burn several of its spars to maintain a head of steam and arrived in New York with only one sack of coal left in its bunker. The *Great Western* arrived early the next morning with ample coal to spare. It would dominate transatlantic transport, cutting eighteen days off the crossing time.

Brunel's next project was to design a steamship even larger than the *Great Western*, and if he was to build one bigger it had to have an iron hull. The world's merchant ships had wooden hulls and many people pointed out that iron couldn't float. The *Great Britain*'s iron plates were riveted together and a workman was accidentally riveted into a sealed space. His skeleton wasn't found until the ship was dismantled.

A major problem on long voyages was that sea water had

to be used to generate steam and every four or five days the boiler had to be shut down and descaled. An Englishman named Samuel Hall invented a device that continually reconstituted fresh water from steam and solved the problem. The *Great Britain* was equipped with this innovation.

Paddlewheels were inefficient because at any time almost all the paddles were out of the water. A Kent farmer named Sir Francis Pettit Smith, whose hobby was making model boats, devised a screw propeller and was so pleased by its performance that he patented it for use in full-sized ships. Shortly afterwards Swedish engineer John Ericsson fitted a propeller of his own design into a ship and proved that it worked well. It took seven years for the Admiralty to consider its merits. It staged a tug-of-war between two frigates, identical except that one was powered by paddlewheels and the other by a single screw propeller. The latter won hands down.

One innovation that Brunel didn't adopt was the brainchild of Sir Henry Bessemer, who invented the Bessemer converter, transforming the production of steel. Sir Henry had to make several business trips across the Channel to France. He didn't enjoy the experience because: 'Few persons have suffered more severely than I have from sea-sickness.' To conquer this malaise he had a novel idea of a saloon suspended from a pivot and weighted down beneath. The idea was that however the ship rolled, the saloon would remain vertical. To show that it worked he built a full-sized replica in his garden with a steam engine to 'rock the boat'. The SS *Bessemer* was built and sent to sea. The saloon proved to be a swinging, swaying ride from a sadistic no-fun fair. The ship was also almost impossible to steer and had a penchant for bumping into piers.

The *Great Britain* was launched in 1843 and successfully crossed the Atlantic for three years until it ran aground on the Irish coast. It sat on the rocks all through the winter storms and was undamaged, whereas a wooden ship would have been reduced to spars and splinters. The loss of income and subsequent refurbishment bankrupted the company. She was returned to active duty until she broke down in the Falkland Islands where she became a floating warehouse for forty-seven years until she was abandoned in what became a graveyard of decaying clippers.

Brunel's enthusiasm was undiminished and he raised funds to build a ship that was two-and-a-half times bigger than the *Great Britain*. It was the biggest movable object that had ever been built. This leviathan had a double-skinned iron hull and was designed to carry four thousand passengers in luxury non-stop to Australia. Its mighty engines slashed the previous record time by two thirds.

Brunel died of a stroke a week after its launch and never knew that it would be a commercial failure. Nor did he anticipate that it would play a vital role in one of the most important ventures of the nineteenth century.

Brunel was a tiny man with a top hat almost as tall as a ship's funnel. But few could match his stature as an engineer or his vision and ambition. As well as railways and the beautiful Clifton Suspension Bridge in Bristol, he built wonderful ships decades ahead of their time. In 1970 the *Great Britain* was rescued from her cemetery in the Falklands and was towed to Bristol, where she now rests refurbished and revived as a glorious example of his genius.

Brunel also laid the foundation for Britain's leading role in transatlantic transportation. British liners almost monopolised the coveted Blue Riband awarded for the fastest crossing. The *Persia* held it for seven years, the *Mauretania* for twenty until surpassed by the *Queen Mary*, which took less than four days. However, the *Queen Mary* was travelling under an assumed name. When she was ready for launching, *Cunard*'s director Sir Thomas Royden asked King George V if he might name the new liner after England's greatest Queen. The King agreed and said that his wife would be delighted. Royden failed to inform his Majesty that the Queen he had in mind was Victoria. Needless to say, that name never graced the vessel.

IRONCLADS, DREADNOUGHTS AND SUBMERSIBLES

'Let us put on armour'
– **Romans, 13:12**

The oceans were not just the highway for trade and exploration. Nations viewed the sea as a stage on which to display their might and a battlefield on which to enforce their will. Special ships were required and warships became the largest fighting machines of their time.

Who rules the waves?

Naval superiority was at the heart of Britain's power. The oak forests of England were stripped bare to make the wooden walls that defended its shores and intimidated rivals. Nelson's ship, HMS *Victory*, was an enormous fort with three storeys of thirty cannons on each side. Every three minutes it could deliver a withering broadside of cannonballs that could burst through a hull 3 ft (1 m) thick to create a blizzard of deadly splinters.

Although warships were armoured with wood the stern was often a mosaic of ornate picture windows. It was as if they had built a battle tank and got a glazier to finish it off at the rear because glass was cheaper than oak. At the battle of Trafalgar in 1805 the *Royal Sovereign* cut through the Spanish line and passed behind their flagship, firing each cannon in sequence. The cannonballs passed through the stern windows and travelled the entire length of the ship. Within one minute it was demolished.

The design of wooden warships had remained the same for three hundred years. They were mobile gun platforms designed to engage the enemy at close quarters. But in the nineteenth century cannons were replaced by ever bigger guns that could hurl larger-calibre shells several miles.

Wooden-hulled ships were now at much greater risk. In 1859 the French wooden warship *La Gloire* was partially plated with iron. The British Admiralty responded by re-inforcing a new iron-hulled frigate with extra plating amidships. Both had a steam engine but were fully rigged sailing ships. Masts and inflammable sails cluttered warships up until the early 1880s. One wonders how many sails went up in flames with a spark from the funnel.

From now on sea battles would be carried out at long range yet for several years the British continued to fit a ram at the prow. It was never used in anger, although HMS *Victoria* did manage to sink another British battleship by accidentally ramming her.

The first fully armoured 'ironclads' emerged during the American Civil War of 1861–5. The Union navy blockaded southern ports to curb the South's main income, from the export of cotton. The Confederate navy was no match for Union ships, so it commissioned an entirely new type of warship. A wooden-hulled steam frigate, the *Merrimack*, was transformed into the *Virginia*, a floating barn roof covered with 4 in (10 cm) of iron plating. She immediately took on three Union ships and her ten cannons disabled two of them and sank the third. On hearing this, the historian Henry Adams declared: 'Almost a week ago the British discovered that their whole wooden navy was useless.'

The Union navy was determined to repay this embarrassment and offered $2,000 for an ironclad to match the *Virginia*. An opportunist proffered plans, took the advance and was never seen again. Subsequently John Ericsson (who had invented the screw propeller) built a well-armoured ironclad. The *Monitor* was a motorised barge whose iron-plated deck was just above the water level with a revolving gun turret amidships. Unlike all other warships, it had a tiny profile to aim at. The two ironclads battered each other with shells for four hours at close range, slugging it out like heavyweight boxers. Neither could breach the other's armour and eventually they disengaged and limped back to port with their deafened crews.

In Europe, ships' guns now had rifled barrels to make the shell spin, which greatly increased their accuracy. William Armstrong's Newcastle factory produced a gun that could fire a 12-in (30 cm) calibre shell weighing 600 lb (270 kg) with a good chance of hitting an enemy ship 10 miles (16 km) away.

The response was thicker armour. The mighty 'Dreadnoughts' had 17 in (43 cm) of teak covered with 24 in (60 cm) of iron. Not surprisingly, the heavier the battleship the slower it moved. It took a dramatic event to solve the problem.

The turn of turbines

Britain's Charles Parsons came from a technology-savvy family. His father had built the largest reflecting telescope in the world and his mother was a pioneer of photography. He worked in an electric dynamo factory just outside Newcastle. A steam engine powered the dynamo via a belt drive, but surely the steam could drive it directly. Parsons established an engineering company and invented the steam turbine. It consisted of a series of vaned wheels on a shaft. High-pressure steam rushed through the vanes, causing them to spin the shaft at eighteen thousand revolutions per minute. The spin was transmitted to a dynamo to generate electricity. Parsons sold the turbines to local power stations and soon the whole of Newcastle was lit by turbines, as almost all cities are today.

He also thought it would be an admirable way to turn a ship's propellers, so he built a 100-ft (30 m) long boat, the *Turbinia*, which could generate 2,000 horsepower. As usual, the Admiralty was not interested. In 1887 Queen Victoria's Diamond Jubilee was celebrated by a naval revue at Spithead. As the majestic ships of the line passed at full speed they looked impressive. From nowhere out zipped the *Turbinia* at 40 mph (64 kph), darting between the ships and leaving them in her wake. It was as if a swallow were taunting a flock of turkeys. Overnight all the navies of the world were interested in turbine propulsion.

The unseen enemy

In 1905 an adviser to President Theodore Roosevelt declared that America's mighty fleet of battleships would guarantee universal peace. It didn't because the Dreadnoughts were not as impregnable as he imagined. The greatest threat to warships in future was not from other battleships, but a small craft that crept up unseen and struck without warning – the stealthy submarine.

It is easier to design a submarine than to build one that works, and it is far less dangerous to build one than to go down in it. Leonardo da Vinci refused to reveal the details of an underwater boat he envisaged because he feared that others might use it for evil purposes.

In 1578 English mathematician, scientist and gunner William Bourne designed 'a shippe that may go under the water unto the bottom and so come up again at your pleasure'. He had grasped the essentials, but sinking is easy: it's the return to the surface that's most appreciated by the crew. His hypothetical submersible was equipped with leather bags pleated like a concertina. When these were compressed flat by a screw press, no water could enter from outside, but if the press were eased water would rush in, making the boat heavier and so causing it to descend. Screwing down would push the water out and hopefully the vessel would ascend. However, the vessel's hull was a wooden frame with only a leather overcoat to keep the water out. Fortunately, it was never built, thus saving several lives.

The first working submarine came fifty-two years later. Its designer was Cornelius van Drebbel, a Dutch mathematician who came to England to tutor King James I's children. When he wasn't tutoring he was advising the Royal Navy on explosives and running a pub. In his spare time he invented the thermostat to keep a furnace at a constant temperature by regulating the air flow. But by 1620, just as the Pilgrim Fathers set sail in the *Mayflower*, he created a wondrous 'submersible galley'.

Drebbel's friend the English physicist Robert Boyle gave a brief description of the submersible in his book *New Experiments*. It was made of wood and was rowed by twelve men both on the surface and underwater at up to 4 mph (6.4 kph). Sinking and rising were achieved with the use of pigskin bags much as Bourne had described. The submersible was taken down to a depth of 12 ft (3.7 m) without mishap. Ben Jonson wrote that Drebbel 'hath made the Hollanders an invisible eel'.

The weakest points were where the oars came out of the hull. In 2002 a two-man version of Drebbel's boat was built using the materials and methods available in the early seventeenth century. It was found that a watertight seal could be secured if the oar went through a wooden cylinder that pivoted in an oar slot packed with greased hemp, and on the outside the emerging oars were enclosed in a heavily greased cuff. The oars didn't need to be feathered – twisted so that the blades were horizontal when out of the water, to reduce wind resistance – because the leather blades folded like an umbrella on the back stroke and opened on the power stroke.

Drebbel's 'submarine displays' on the River Thames were the talk of London. Up to sixteen passengers at a time were taken on excursions, propelled by twelve oarsmen. They must have been packed like dates in a wooden box. The vessel was said to have stayed underwater for three hours. This was possible because it had a tube and bellows to suck in fresh air when the submarine was just awash. It is also likely that Drebbel, who was an alchemist, had invented a way to replenish the air while the sub was fully submerged. Boyle, a fellow alchemist, wrote that Drebbel 'had Chymicall liquor which he accounted the chiefe secret of his Submarine Navigation. For when from time to time he perceiv'd that the purer part of the air was consumed . . . he would by unstopping a vessel full of his liquor . . . make it again, for a good while, fit for respiration.'

Drebbel's *Treatise on the Elements of Nature* states that heating potassium nitrate gives off 'the nature of life itself'. It does indeed generate oxygen and there can be little doubt that he used oxygen generated in the lab to refresh the air in the submarine. This was 140 years before oxygen was identified and named. However, preventing a build-up of carbon dioxide is just as important as ensuring the supply of oxygen. Caustic soda is excellent at 'scrubbing out' carbon dioxide from the air. It is not inconceivable that Drebbel knew this too.

In 1653 a Frenchman named Monsieur de Son made great claims for the performance of his twin-hulled submersible catamaran armed with a huge iron ram: 'The perfect forme of the Strange Ship . . . doeth undertake in one day to destroy a honderd ships, and goe from Rotterdam to London and back againe in one day . . . and to run as swift as a bird can fly.' The bird in question was a dodo. The ship was propelled by two large paddlewheels powered by clockwork and it went no faster than the hour hand.

Bishop John Wilkins, who was Oliver Cromwell's brother-in-law and a friend of Drebbel, coined the word 'submarine' and shrewdly anticipated its military potential: 'A man may thus go to any coast of the world invisibly . . . It may be of very great advantage against a Navy of enemies, who by this means may be undermined in the water and blown up.'

The first to put this advantage into practice was an

American, David Bushnell. In 1775 he petitioned George Washington for support to build a submersible to attack British ships. With the War of Independence imminent Washington granted the funds. Although Britain played no part in the development of submarines, a dislike of the British was, as we shall see, a great stimulus for others to pursue it.

Bushnell's *Turtle* looked more like a giant acorn than a submarine. It was a large barrel topped by a turret with a tiny peephole. Inside, it was a sophisticated machine for its time. The lone pilot sat amid numerous controls and operating them was like being an extreme one-man band. There were two hand-cranked screws to pull the vessel forward and the rudder bar was clamped under his arm. The *Turtle* was the first submersible to have proper buoyancy tanks like a modern submarine and the water level in the tanks was adjusted with two pistons worked by foot pedals. The various dials were illuminated by the fluorescent glow of decaying wood. Attached to the outside of the hull was a mine containing 150 lb (68 kg) of gunpowder. The idea was to drill a wood screw into the hull of the target ship and attach the mine, which was detonated by a clockwork time switch. Wisely, Bushnell got someone else to pilot his machine.

The British fleet was anchored ostentatiously off Long Island knowing that it was invulnerable. The *Turtle* was moored in a bay nearby. As fate has a wicked sense of humour the British chose the very same bay to land twenty thousand troops. The *Turtle* had to be quickly concealed, although what looked like a large barrel would probably not have aroused much suspicion.

The next night the *Turtle* drifted downstream towards Admiral Howe's flagship, HMS *Eagle*. The current carried it too far and after two hours of hard cranking it had nestled beneath the *Eagle* and the pilot set to work drilling into the hull. Fate smirked again: the hull was sheathed with copper to prevent fouling by boring organisms such as ship worms and, as it turned out, turtles. The pilot surfaced and cranked away as fast as he could, but in the early light the *Turtle* was spotted and a longboat gave chase. It was the hare and he the tortoise, so he released the mine and its clock began to tick. The pursuing sailors didn't like the look of it and turned about. They were pleased they had when there was an almighty explosion.

That at least is the legend. Experts think it is unlikely that an exhausted man using such crude devices could have manoeuvred the *Turtle* beneath the ship and kept it there for the stated time without being overcome by carbon dioxide. Apparently the log of the *Eagle* makes no mention of an incident that night, not even the explosion.

There were two further attempts to fix a mine to a ship but without success. After his sponsor lost faith, Bushnell lost hope. He shortened his name to Bush and became a country doctor.

Before he became a successful ferry boat manufacturer, Robert Fulton was an inventor of boats that could convert into submersibles. He built the *Nautilus* in 1798. It had a weird sail and the mast was hinged at the base so that it could fold away into a channel on the deck. The streamlined hull had adjustable hydroplanes enabling it to dive rather than just sink. Underwater it was powered by four crewmen cranking a propeller that Fulton called 'the Flier', which whooshed the *Nautilus* along, though at only 2 mph (3.2 kph). He claimed that it could stay submerged for six hours or more because it was fitted with a large tank of compressed air.

Like the *Turtle*, the *Nautilus* carried a mine to be attached to a ship's hull, and in a demonstration it sank a single-masted sloop. Fulton met Napoleon Bonaparte and regaled him with images of submarines blockading the Thames estuary and negating Britain's sea power. Napoleon was naturally beguiled by such a scenario, but Fulton refused to hand over the plans and demanded a commission in the French navy. France's admirals disliked his submersible because it was so effective: 'Citizen Fulton's submarine boat is a terrible means of destruction because it attacks in silence in a manner which is nearly unassailable.'

The British Admiralty got wind of Fulton's impressive demonstration and his plan to humble Britannia. In a confidential paper to the Lords of the Admiralty it was admitted that 'Ships in the port of London are liable to be destroyed with ease.' The solution was to buy off Fulton and his invention and thus deny it to the French. So Fulton abandoned his allegiance to France and took the first boat to London. He chilled the Prime Minister, William Pitt the Younger, by boasting that his submarine would 'lead to the total annihilation of the existing system of marine warfare', the very system in which the British were expert. The American demanded £100,000 (around £6 million today) for his services but settled for a monthly salary of £200 (£15,000). The Admiralty set him to manufacture powerful mines that could either float or be moored beneath the waves waiting for an unsuspecting ship, even though an admiral stated that blowing up French frigates with mines 'was not in keeping with the highest traditions of His Majesty's Service'.

Fulton submitted plans for an improved submersible and when the government's deliberations were slow he vented his impatience by writing hectoring letters to ministers and even the Prime Minister. He was exploding more frequently than his mines and losing friends fast. His timing couldn't have been worse. Word came that the French and Spanish fleets had been destroyed at Trafalgar. Nelson had sunk Fulton too. What need did the triumphant Brits have for a piddling submarine?

So Fulton departed for home more disgruntled than ever. In American history he is erroneously celebrated as the inventor of the steam boat, but his pioneering work on submarines is largely forgotten. Although his vessels were far short of perfect, they demonstrated that an undersea raider could destroy a surface ship.

Soon even the most unlikely people were inventing submarines. Captain Johnson, a notorious British spy and smuggler, planned to smuggle Napoleon from St Helena in a clockwork submarine. He pocketed the £40,000 fee, but a week later Napoleon died. Johnson built five submarines but as the world had a temporary absence of major conflicts there was no market for new weapons.

Really good at sinking

The American War of Independence was a testbed and a deathbed for submersibles. In 1863 a Confederate captain whose parents thought Horace Hunley was a mellifluous name, designed a submarine to sink the Union fleet. They were classed as 'Davids' because their prey were Goliaths. They were built from old iron boilers salvaged from abandoned river boats and their weapon was a mine on the end of a long pole at the front of the submarine. With this surprise lollipop they hoped to lick the enemy. They were the deadliest attack subs ever built, but mainly for their crew. In failed attempts to attack a ship four Davids sank with a total loss of thirty-three crewmen. A fifth sortie was successful in so far as it blew up a Union ironclad. Sadly, the blast also sank the submarine and nine more crewmen. No American submarine could boast of sinking a ship for another seventy-seven years.

The next setback for the US Navy was buying Oliver Halstead's submarine, the *Intelligent Whale*. It was not the smartest cetacean in the sea. It had an air lock to enable the crew to escape. They would have been well advised to have used it, for thirty-nine men were drowned during its trials. Even so, Halstead was keen to persevere, but luckily he was shot dead by his mistress's other lover.

Meanwhile in France Captain Bourgeois and engineer Brun produced an enormous iron cigar named *Le Plongeur* (the diver). It was 140 ft (43 m) long and featured several innovations, including watertight partitions and a piston engine powered by compressed air. From then on no one would have to hand-crank a propeller. The only thing *Le Plongeur* lacked was discipline. When submerged it would suddenly rear up unbidden or plunge down on a whim. The navy thought it safer to use it as a large water tank on land.

In England the Reverend George Garrett was doing the Lord's bidding by building a submarine to sink ships. His first attempt was described in the press as 'very nearly successful'. Unfortunately, for a vessel that takes men under the sea nearly successful isn't good enough. A steam engine propelled the sub on the surface but the fire was extinguished as soon as it submerged and steam stored under pressure powered it underwater. *Resurgam* ('I shall rise again'), as it was called, promptly sank while on tow to its first outing and didn't rise again for over a hundred years, until a fishing boat snagged its net on the sub in 1985. Garrett went on to co-design other innovative submarines and, as clergymen do, he fought in the Spanish–American war.

The admirals of great fleets considered the submarine to be of interest only to those countries with a puny navy wishing to irritate the great powers. How right they were. John Holland was an Irish nationalist resentful of British indifference to the suffering of Ireland during the great famine. A frail, myopic schoolteacher, he was an unlikely threat. With his doleful expression and a bowler hat permanently glued to his head he might pass for a funeral director, but whose funeral would it be?

In 1875 Holland emigrated to America and began designing submarines. The US Navy dismissed his ideas as the 'fantastic scheme of a civilian landsman'. After the previous debacles with Hunley and Halstead they considered submarines to be merely expensive iron coffins.

The angry Irishman

Holland mixed with the leaders of the Fenian movement in his adopted country, who were ever ready to embarrass the

British Empire. They had dispatched troops to invade Canada, but they got lost and hungry and turned around. The leaders rather fancied having a submarine at their disposal and their 'skirmishing fund' raised money to build it. Holland's first sub was launched in 1878. It skimmed down the slipway and kept on going until it sat on the bed of the river. Although there were frequent disputes over the finance, its successor, *Holland II*, was finished three years later. It had a petrol engine for propulsion when on the surface and a large reserve of negative buoyancy so that it could surface rapidly in an emergency.

The British chargé d'affaires in Washington demanded that the US government should seize what the papers had christened 'the *Fenian Ram*', but without success. An engineer took the *Ram* on a joy ride and suddenly the waves poured into the open hatch. The engineer below was blown out of the sub by the up-rushing air. No sooner had the *Ram* been salvaged and dried out than three other Fenians, after a sip or two of Guinness no doubt, stole it and careered around New Haven harbour at night towing a one-man sub that Holland had also invented. The little sub was swamped, never to be seen again, and the *Ram* caused so much havoc that the harbourmaster banned it. Thereafter it was grounded and its engine filched. A despairing Holland said: 'I'll let her rot on their hands.'

In 1888 he won a competition to design a submarine for the US Navy, but when it was only half built the funding was withdrawn. Five years later he resumed where he had left off, but it became obvious that the amendments made by the Navy Board would fatally compromise the vessel. But in secret and with his own money he had been building another submarine. The *Plunger* was the first steel-hulled submersible and became the first working submarine in the US Navy.

Holland founded the Electric Boat Company (now subsumed in General Dynamics) and became a leading submarine manufacturer. The design was not yet perfect: for example, the enormous periscope was fine if you were looking forward, but looking astern the image was upside down and when looking to the beam of the sub everything was on its side.

The Admiralty awakes

In 1901 the Royal Navy bought five of the latest Holland submarines. Having taken no interest whatsoever in the development of underwater vessels, it found itself with a flotilla of the most up-to-date submarines in the world. Ironically, Holland's vessels, born out of a hatred of Britain, would now augment its naval power.

That's not to say that the admirals had overcome their prejudice against submerged craft. As late as 1910 Rear Admiral Wilson branded submarines as 'Under-handed, unfair and damned un-English. They'll never be any use in war and I'll tell you why . . . we intend to treat all submarines as pirate vessels . . . and we'll hang all the crews.' It was the German U-boats that brought Britain almost to her knees in both World Wars. Dashed unfair, but bloody effective.

Well, effective if they weren't British K Class submarines of 1917. Both K2 and K13 sank on their maiden voyage. K1 collided with K4 and sank. K3 dived inexplicably and when coaxed back to the surface was rammed by K6. K5 foundered in the Bay of Biscay, but K15 took no chances and sank in the safety of Portsmouth harbour. On an exercise to integrate the subs with surface ships K4 ran aground after ramming and sinking K6, while K17 collided with K7 and was then sunk by a cruiser. K22's helm jammed and she was permanently disabled. The exercise is fondly known as the 'Battle of May Island'.

The early pioneers of submarine design were hampered because they had neither an efficient means to propel the sub underwater, for engines gave off noxious fumes, nor a way to strike at the enemy from a distance while submerged. The advanced Holland submarines had a non-polluting electric motor powered by sixty batteries charged by the internal-combustion engine while cruising on the surface, and they carried five torpedoes.

The torpedo terror

Robert Whitehead was an engineer who specialised in designing silk-weaving machines and drains. Then in 1866 he produced an explosive torpedo powered by compressed air driving a piston engine that turned twin propellers. 'Torpedo' is the name of an electric ray fish that can give a mighty shock of electricity to kill its prey. Initially Whitehead's torpedo had a range of 600 ft (183 m) and travelled at just over 6 mph (9.7 kph), although both these figures were soon vastly increased. As it couldn't be steered it could hit only stationary ships and it had an annoying habit of missing the target. It was not until 1895 that an Austrian, Ludwig Obry, invented the gyroscope, an inertial guidance device that was ideal for keeping a torpedo on a straight course. Whitehead invented a device to keep the torpedo at a pre-determined depth.

Ships had no defence against torpedoes except to pray they would miss. So in 1885 the British Navy developed a new type of light, fast ship called a torpedo boat destroyer, later abbreviated to 'destroyer'. They were armed with guns and torpedoes launched from the tubes, designed by Whitehead, to sink ships before they had time to release *their* torpedoes.

To hit fast-moving ships there had to be a more sophisticated way of guiding a torpedo to its target. Wireless waves provided the answer, but they were readily intercepted, so the torpedo could be deflected. The solution to this problem came from a very unusual source.

Hedwig Kiesler was an Austrian actress. In 1933 she caused a sensation by gambolling naked in a film whose title she fully justified: *Extase* (*Ecstasy*). The film was banned in several countries but a few copies found their way to Hollywood, where they were viewed at private parties. Movie moguls decided Kiesler was someone they just had to have – they might even put her in a film. 'Hedwig' became 'Hedy', and 'Kiesler' had to go too because Cecil B. DeMille thought it sounded like a rude word for backside. So the stunningly beautiful Hedy Lamarr was born. In the sword-and-sandals epic *Samson and Delilah* she was gorgeously wooden. She never claimed to be a great actress, and said that to be glamorous you just had to 'stand still and look stupid'. But she was anything but stupid.

Hedy had been married to an Austrian arms dealer and she was privy to his discussions with Nazi officers on the newest weapons. She heard of the problems with torpedo guidance systems. Wireless waves (of which more later) would be ideal if only the signals weren't so susceptible to jamming. She had an idea about how to prevent the jamming of wireless waves. For practical advice she turned to her friend George Antheil, not a submarine expert but a composer. George shunned instruments in favour of car horns, rattles and whizzing propellers. More significantly he had conducted a chorus of four synchronised player-pianos while the audience went wild and threw things.

Hedy and George realised that both the wireless transmitter and the receiver had to be coordinated in some way so that the signal could switch from one wavelength to another and the receiver would change to that frequency at the same moment. To a potential jammer the variations would be random and therefore immune to interference. When trying to explain this to the US Navy George suggested that both the transmitter and the torpedo could have small paper rolls punched with identical pattern of random holes (a miniature version of the rolls in a player-piano) running in synchrony. The frequency would be determined by the sequence and position of the holes. The idea was rejected and torpedoes continued to go astray.

It took decades before 'frequency hopping spread spectrum', as it came to be called, was used to prevent the jamming of radio waves. What's more, it's now vital to prevent mobile-phone conversations interfering with each other.

Torpedoes fired from submarines were one of the most effective weapons ever invented. In the First World War over twelve million tonnes (11.8 million tons) of shipping were sunk by torpedoes, while the Admiralty was considering the effectiveness of seagulls as a weapon. In the Second World War the figure was more than twenty million tonnes (19.7 million tons), including more than two hundred warships.

MORE COMPLICATED WAYS
TO DROWN

'He who would search for pearls
must dive below'
— John Dryden

S ubmarines are designed to submerge, but ships sink by
misadventure. They can be swamped by storms, bitten
by sharp-toothed reefs, stung by torpedoes or led astray by
captains who are asleep or drunk. There are forty thousand
known wrecks around the British Isles alone and many more
lie undiscovered throughout the world's seas. Once ships
descend into the darkness they are lost and forgotten – unless
they contain a valuable cargo.

Treasure hunting below the sea began long before proper
equipment was available, but as Joseph Conrad said: 'There
are few things as powerful as treasure once it fastens itself
on the mind.' Far more treasure ships have sunk than ever
sailed, but some brave men *have* recovered riches.

Emperor Caligula longed for a new orgytorium so that he
might debauch in a novel setting. Caligula was not known for
his frugality: he squandered the entire contents of the Imperial
treasury in a single year. He built two ships over 200 ft (61 m)
long and 70 ft (21 m) wide, to float on Lake Nemi near Rome.

They were encrusted with precious stones and full of statues, wonderful carvings, baths and bordellos. But the emperor soon tired of bouncing around on the lake, so the boats were left to leak and came to rest 60 ft (18 m) down on its bed.

In 1535 engineer Francesco de Marchi was charged with salvaging the treasures. He hired a diver named Guglielmo de Lorena, who wore his own invention – a wooden box with an open bottom that went over his head and reached down to his waist, leaving his arms free to collect whatever he found. He peered through a small glass window and breathed the air trapped inside the box, but had to return to the surface frequently to replenish the air. It was the first use of diving equipment in a salvage operation.

Over the ensuing centuries there were repeated raids on the two wrecks, with the booty going to the highest bidder. Even the decking was carved into walking sticks that sold well. The ships were too fragile to be raised, but in 1928 the Italian dictator Mussolini, another spendthrift, ordered that the lake should be drained to uncover the wrecks. It took three years and the ships were conserved and housed in a custom-made museum, until fire ravaged the museum and both vessels were destroyed.

When ships founder at sea, draining the ocean is not an option, so divers have to go down. The best appliance available from the sixteenth until the nineteenth century was the diving bell. This is simply a watertight container inverted and open at the bottom. If you take a tumbler and hold it upside down in a bowl of water you have a model diving bell. The diver breathes the air trapped inside, so the bigger the bell the better. However, the early versions were 'worn' by an individual. Signor Lorena's was a wooden box; others resembled a diver trapped in an hourglass within a wooden cage, or the man in the iron flask, or an escaped lunatic wearing a misplaced kettle. Many designs have the diver breathing through a long tube to the surface. This doesn't work because a diver can't suck down air against the water pressure even when he is only 6 ft (1.8 m) beneath the surface. One diver had his head stuck into the base of an enormous wooden pipe that reached up all the way to the surface. No matter how loudly the boat crew shouted, 'Are you all right?', he never replied.

In many illustrations the diver's countenance bears the sad resignation of one who fears the worst. A drawing from 1664 of a man strapped inside a bell made of leather shows the water coming up to his ankles and at that depth it was probably safe. In another depiction the diver seems to have purloined a real bell with an over-developed clapper. We can only see his shins protruding beneath the bell, but between his legs there is an enormous cannonball designed to send him to the sea floor and ensure he stays there.

In 1640 a ship exploded and blocked access to Charleston harbour, in what would become South Carolina. Edward Bendall built two wooden bells to help him clear the wreckage. He also recovered a cannon that was stuffed with rope yarn. To celebrate the end of his task he fired the weapon and a hailstorm of gold and silver coins shot out of the barrel and disappeared into the sea.

Hell's bells

An Englishman, Edmond Halley, designed the first large, safe and successful diving bell in 1690. This was before he became Astronomer Royal and the predictor of a comet's return.

He became interested in diving when a frigate sank in Pagham harbour in West Sussex with a valuable cargo that was being clandestinely imported by high-ranking people. If it were to be salvaged, a discreet man must be hired. That man was Halley with his diving bell and the cargo was retrieved from a depth of 60 ft (18 m). Exactly what was retrieved, or how much of the total cargo, remains a mystery.

Halley described his bell as a large 'truncated cone' with a glass window in the roof to let in light. The wooden walls were lined with lead to ensure it sank. He calculated how long the air inside the bell would last with a nervously breathing man inside. He also mentioned its most disconcerting feature: 'as it descends lower, [the air] does contract itself according to the weight of the Water that compresses it; so as at thirty-three Foot [10 m] deep or thereabouts, the Bell will be half full of Water.' The diver sitting on a bench inside came to appreciate that it was the pressure of the insubstantial air within the bell that kept the water at bay. Halley himself was lowered to a depth of 54 ft (16.5 m), where he remained 'without much inconvenience'.

It was possible for five divers to stay down for an hour and a half because Halley had devised a method of supplementing the air supply. A small tap at the top of the bell vented out the spent air and fresh air was lowered down in lead-lined barrels. The bell was primarily a means to reach the sea floor and act as a workstation for divers who left the bell and sometimes trudged 'a good distance away'. This was achieved by the diver wearing a leather hood weighted to stay on his shoulders and connected to the bell by a long, flexible breathing tube. To brace the tube against the pressure, Halley used a spiral of brass wire covered with thin leather and 'dipt into a mixture of Oyl and Bees-Wax hot [which] made it impenetrable to water'. It was finished off with 'several folds of guts' that were allowed to dry and shrink around the tube, then a coat of paint and more leather. It was a lifeline for the diver in two ways. It provided them with air and was 'a clew to direct him back again, when he would return to the bell'. Later Halley replaced the hood with a lead helmet with two eye windows. He also invented the diver's weight belt and 'clogs of lead for the feet' to enable the diver to hold his position in a strong current. These became part of the diver's standard gear.

Halley invented special tools for divers because hammers and crowbars didn't work well underwater. Particularly useful were levered tongs for lifting objects without bending over. If the diver leaned down lower than the level of the water in the bell, his air hose sucked back air instead of blowing it down to him. Halley also devised a depth gauge, not unlike the one invented by Robert Hooke in 1663.

The dangling coffin

There was an alternative apparatus in which the diver was protected from the water outside and breathed air at atmospheric pressure so that he was not at risk from either 'the squeeze' or 'the bends'. It was invented in 1715 by an impoverished Englishman named John Lethbridge. Like the old woman who lived in a shoe, he had so many children he didn't know what to do until he came up with an 'extraordinary method to retrieve my misfortunes; and was prepossessed that it might be practical to contrive a machine to recover wrecks lost in the sea'.

Lethbridge tested his idea in a flooded trench dug in the garden by climbing into a large wine cask, securing the lid and submerging. There was sufficient trapped air for him to spend over half an hour underwater in such comfort as an unfurnished barrel could provide. Encouraged by this, he got a cooper to make him an elongated barrel 6 ft (1.8 m) in length and reinforced with metal bands. There was a glass viewing port and two holes fitted with leather sleeves and gauntlets.

The weighted barrel was lowered from a boat to just above the sea floor. Lethbridge lay chest-down, peering through the window while picking up objects and putting them into a basket to be hauled up to the surface. With his hands confined to the gauntlets he had to wipe the condensation from the window with his nose. When he had almost exhausted the air inside the barrel he was winched up to the boat to vent the foul air and have it replaced by means of a bellows pump. Sometimes he almost left it too late and nearly asphyxiated. He wrote: 'I have been to ten fathoms [60 ft/18 m] deep many more than a hundred times and have been to twelve fathoms [72 ft/22 m], but with great difficulty because of the pressure.'

The 'Silver Fisherman' as he became known was commissioned by the Dutch East India Company to recover bullion from wrecks all over the world. He was the most successful treasure hunter of the eighteenth century and it is estimated that he earned a sum today equivalent to £8 million.

Major operations

Variations of Halley's big bell were still the mainstay of salvage operations a century later. In 1818 Charles Babbage, the inventor of the first-ever computer, recorded his experience on descending in a metal diving bell: 'A pain begins to be felt in the ears arising from the increased external pressure; this may be removed by closing the nostrils and mouth and attempting to force air through the ears. As soon as the equilibrium is established the pain ceases, but recommences almost immediately by the continuance of the descent . . . Signals are communicated by the workmen striking against the side of the bell with a hammer.' The most used message was given the fewest number of strikes. One strike was a plea for 'More air!'

The techniques of underwater salvage were honed on the wreck of the *Royal George*. In 1782 this ageing man-of-war was resting at anchor in Portsmouth harbour. Before it sailed the crew were entertaining almost three hundred guests on board – a mix of sailors' wives and female acquaintances of the 'most depraved character'. The ship had been heeled over to expose a hole in the hull so that it could be patched. Rum has made many a mariner unstable; loading a large shipment of rum on to the *Royal George* caused her to lean too far and capsize. William Cowper penned a long and turgid ballad to the victims who thankfully were spared from having to listen to it.

This sunken mausoleum was an obstacle to shipping but was too heavy to lift, so the rights to salvage the wreck were granted to a young man called Charles Deane. He had been a caulker, sealing leaks in planking. Then in 1823 he patented a smoke hood to be worn when fighting a fire. The equipment consisted of a lightweight, three-windowed leather helmet riveted to a leather tunic. The helmet was continually flushed with fresh air from a hose attached to a bellows pump. Although it was an excellent invention no one was interested.

With the help of his brother John, Charles converted the fire helmet into diving gear. This was an open-helmet apparatus similar to Halley's invention a century earlier – a case of reinvention. To commercialise the equipment Augustus Siebe was asked to manufacture several helmets to Charles Deane's design. Siebe had been a German artillery officer at Waterloo who became a gunsmith and instrument maker in London. John Deane wrote the very first diving manual to instruct purchasers. It claimed: 'A person equipt with this Apparatus, being enabled to descend to considerable depth, from twenty, probably thirty fathoms . . . and to remain several hours . . . is freely able to traverse the bottom of the sea, and to search out the hidden treasures of the deep.'

The Deane brothers brought up twenty-two valuable cannons from the *Royal George*. John Deane was asked by a fisherman if he would free his net that was snagged on the seabed. The obstacle was the wreck of Henry VIII's prize battleship, the *Mary Rose*. She had been lost for three hundred years and would remain enshrouded in silt until she was dramatically raised to the surface live on TV in 1983.

In 1830 an experienced bell diver suggested to Charles Deane that the air supply should come from one of the new force pumps instead of inefficient bellows. At the same time Thomas Hancock invented a flexible rubberised 'Aquatic dress for walking across Deep Water . . . made so as to perfectly Exclude the Water'. The combination of the close-fitting diving suit and Deane's improved helmet was a significant advance.

Charles was devastated when his contract to salvage the *Royal George* was not renewed. Moreover, his inventions had not been the commercial success he had hoped for. He never dived again, whereas his brother John's career thrived. During the Crimean War he was commissioned as an explosives and diving expert.

Engineer George Edwards invented a flange on the rim of the diving helmet that clamped on to the rubber suit. This meant that the diver's helmet and suit became one. Edwards gave Siebe a free hand to use his invention and Siebe took full advantage. He even asked Edwards for his detailed plans to save him the expense and time of re-inventing it.

After Siebe died the company he had founded trumpeted him as the inventor of the 'closed' diving dress, which was Charles Deane's outfit modified with Edwards's flange seal. Siebe had merely manufactured these after Deane's patent expired. Deane always feared that his name would be eclipsed by Siebe's and he was right. Siebe Gorman became the world's foremost manufacturers of diving equipment and the closed diving dress became the standard gear for underwater workers until the mid-twentieth century.

The salvage of the *Royal George* was renewed by explosives expert Colonel Charles Pasley and his team of sappers. They systematically blew up parts of the ship to gain access to its interior. Health and safety was unheard of, the gunpowder was poured into barrels through a short metal tube and then the tube was *welded* shut. The divers burrowed through the silt all the way under the hull so that a sling could be drawn under it and winched up to retrieve sections of the ship.

The divers often worked in zero visibility when the silt was disturbed, so they relied on 'feel'. One diver encountered something floating just above the sea floor. He ran his hand along a sort of corrugated grating. It was a human rib cage.

The next significant advance in underwater salvage came in 1916 when the Italian battleship the *Leonardo da Vinci* was sabotaged in Taranto harbour and came to rest upside down in mud. General Ferrati, the chief of naval construction, had an ingenious idea. His divers covered every hole in the hull by riveting steel plates over them with a rubber sheet below the plate to ensure a waterproof barrier. When the hull was absolutely airtight, compressed air was pumped into the ship to turn it into a heavyweight balloon. The leviathan rose to the surface like a lazy whale. It was towed into deep water and heavy ballast was loaded on one side, then selected compartments were flooded to increase its instability until it overbalanced and floated the right way up. The mission was a triumph, but it had taken four and a half years to salvage one battleship. Surely raising an entire fleet of warships was out of the question.

At the end of the First World War the German high-seas fleet was interned in Scapa Flow in the Orkney Islands off northern Scotland. The skeleton crews of German sailors could see the lights of Kirkwall, Orkney's capital, but they were confined to their ship and their frustration led to mutiny. On a June day in 1919 a red pennant hoisted on the cruiser *Emden* was the signal for all the crews to open the sea cocks, portholes and watertight doors. One by one almost all fifty-two ships sank beneath the waves, leaving only the tips of a few masts poking above the surface like the remains of a drowned forest.

To Ernest Cox they were an enormous cache of valuable scrap metal. He was an engineer who had made a fortune manufacturing weapons and buying scrapped military equipment and he would risk it all on this salvaging adventure. He purchased the German ships at knockdown prices and now all he had to do was get them to the surface. A huge floating dock was split in two to serve as pontoons that could be sunk beside a ship and secured by slings. When the pontoons were filled with air they lifted the ship to the surface. Within three years Cox's team had retrieved twenty-five destroyers. They became so skilled that latterly they were raising a ship a week.

The monstrous battleships were a different matter. Some weighed over 28,000 tonnes (27,500 tons), more than all the destroyers put together, so Cox used Ferrati's method. All the holes had to be patched by divers in standard diving dress. One battleship required over eight hundred patches, one as large as 40 ft (12 m) by 21 ft (6.4 m).

When the battleship the *Hindenburg* was being drained, a diver checking for leaks got the seat of his rubber suit and a significant chunk of his rump sucked into the ship's sea cock. The suction was so great that fellow divers couldn't release him. There was no alternative but to partially re-flood the hull to reduce the suction. The sore-bottomed diver's explanation was that he had been sent down to seal any leaks, and that was exactly what he did.

To gain access to the ship's interior the team constructed

air locks in vertical pipes like factory chimneys reaching from the exterior of the hull to above the sea's surface. Some were 90 ft (27 m) tall and wide enough for a man to climb down on an interminable ladder. At the bottom of the pipe was a hatch giving access into the air-filled hull. Some workmen spent eight hours a day in this dark, eerie, upside-down world. One diver was shaken to encounter the corpse of a sailor still blanketed in his bunk. It turned out to be a dead seal dressed in a sailor's uniform by a fellow diver with a warped sense of humour.

Cox would not get paid until the *Hindenburg* was delivered to the shipyards in Rosyth, near Edinburgh. It had to be towed 280 miles (450 km) with fifteen workmen living on the upturned hull tending the compressed-air pumps, only 4 ft (1.2 m) above sea level. It was like camping on a giant whale and hoping it wouldn't dive. Another of the huge battleships just missed crashing into a pier of the Forth Bridge.

Cox salvaged eight battleships, one raised from 105 ft (32 m) down. Some of the steel was incorporated into the *Queen Mary*. Only seven ships remain on the bottom of Scapa Flow, which is probably the most popular diving site in the British Isles.

Diving in the standard dress is not without risk. So long as the air the diver breathes is at the same pressure as the surrounding water he feels no discomfort, but should the air hose split he is suddenly and terribly exposed. It's fondly called 'the squeeze', but in severe cases the diver's body is rammed into the helmet and the body fluids are sucked up the hose. To prevent this unpleasant surprise Siebe Gorman put a non-return valve in the helmet so that if the hose split the air would simply be cut off, which was not ideal, but better than the squeeze. Another problem was the dreaded 'bends', caused by staying too long at depth and then not taking sufficient time to rid the body of the nitrogen it had absorbed under pressure. A too hasty return to the surface caused the gas to fizz up in the blood and joints, leading to paralysis or worse.

Deep, dark and dangerous

For very deep diving, how could the diver be protected from the crushing pressure? With a suit of armour, of course. The earliest attempt was a copper breastplate and helmet, although even the inventor conceded that the air hose got squashed, restricting the diver's breathing, and his leather-bound limbs were so compressed as to 'stop the circulation of the blood, as some have experienced to their cost'. Obviously partial armour was little better than no armour at all.

The concept of underwater armour inspired inventors to create a variety of impractical outfits that could only be described as a walking (and drowning) concertina. Among these were an outfit that made the diver look like a Michelin man on steroids, and another that had him entirely wrapped in wire to create a living Slinky, although not one who lived very long.

In 1856 L.D. Phillips, an American, encased the diver in a tubby boiler that had articulated arms ending in large 'lobster' claws. It also had a balloon on top and a propeller at the level of his navel – its purpose remains a mystery. A trio of Frenchmen plumped for separate iron plates that fitted together like a jigsaw. It is obvious that every joint is an invitation to the ingress of water, yet the Carmagnolle brothers thought the more joints the better. Their all-enveloping armour boasted twenty-two joints and the helmet had twenty-four peepholes for a diver with only two eyes. Some suits even added articulations where the human body had no joints. The poor divers were knights in leaky armour.

The first inventors to manufacture watertight, articulated joints were the German engineers Hans Neufeldt and Karl Kühnke in 1913. Their massive suits resembled Humpty Dumpty with elephantiasis in every limb. The outfit carried an ample supply of oxygen and was only attached to the surface ship by a steel cable and a telephone line.

The great advantage of completely enclosing the diver was that he could penetrate much deeper than the diver in standard dress. But the pressure increased the friction on the joints to such an extent that he was as useless as an arthritic scarecrow when 300 ft (91 m) down. He might as well have been in a watertight chamber with no arms and legs protruding.

In 1912 Robert Davis, the manager of Siebe Gorman, invented the Deep Sea Observation Chamber. It was a pressure-proof, one-man cylinder with viewing ports and its own oxygen supply, plus the facility to absorb carbon dioxide from the air. It became a standard feature of underwater salvage operations.

On a calm May evening in 1922 the P&O liner *Egypt* was cruising past the coast of northern France in dense fog. The passengers had just finished their dessert when the *Egypt* suddenly lurched to port and passed the point of no return.

She had been rammed by a much smaller ship whose bows were strengthened for ice-breaking. The *Egypt* sank in deep water, taking with her many tons of gold bricks, silver ingots and sovereigns.

Lloyd's insurers declared the bullion irretrievably lost, but Peter Sandberg, a Swedish engineer, thought that with the right equipment it could be retrieved. He enlisted an Italian salvage company called Sorima (Società Ricuperi Marittimi). The wreck was 396 ft (121 m) down, well beyond the range of standard diving gear. Captain Quaglia of the salvage ship *Artiglio* and his chief diver built their own underwater chamber with a floodlight. To lift debris or treasure they designed remotely controlled grabs and a giant electromagnet as well as a vacuum tube to suck up small objects such as coins.

The man in the observation chamber directed the excavation by giving instructions to the crew. The ship was anchored by six heavy blocks so that its position could be finely adjusted by pulling on one anchor line while slackening that on the opposite side. In this way the grab could be moved to specific places as instructed. Explosive charges were placed in the same manner. Accessing the bullion room required blasting through four decks then clearing out 500 tonnes (490 tons) of metal debris. Often the wreckage in the grab was so heavy that the *Artiglio* heeled over and the sea poured over the deck – the salvage firm had already lost one ship that capsized while lifting. It lost another while salvaging a munitions ship. The chief diver laid a small charge to gain access to the interior of the ship and the *Artiglio* retired to what was the usual safe distance before the chief touched the two electric wires together. It was the last thing he did. The small explosion set off the wreck's cargo of TNT and the *Artiglio* was blown out of the water. (Wrecked munitions ships do not make good neighbours. The *Richard Montgomery* sank sixty years ago in the Thames estuary. Its rusting cargo of unstable bombs amounts to 14,000 tons (14,224 tonnes) of TNT. The explosion of a single bomb could detonate the lot. It would be the largest non-nuclear explosion since Krakatoa blew its top in 1883. That was in Sumatra, the *Richard Montgomery* is only 37 miles (60 km) from the centre of London. To enhance the fireworks display a gas terminal has been built only 1.5 miles (2.5 km) away.)

Quaglia was undeterred and was back the following year with a new ship. The grab brought up the captain's safe from the *Egypt* containing the key to the bullion room. All Quaglia had to do now was to find the lock. Eventually the grab brought up the first gold bricks, but it took three years to recover 7 tonnes (6.9 tons) of gold and almost 40 tonnes (39.4 tons) of silver bars and coins.

Today the observation chamber has been replaced by a remotely controlled camera that transmits images to the mother ship via fibre-optic cables. The cutting and lifting equipment is now far more powerful and more dextrous and can be deployed with great accuracy. The development of immensely strong cables that are weightless underwater has enabled salvage operations to be carried out at depths of 11,500 ft (3,500 m).

Flying underwater

The Frenchman Jacques Cousteau became the most famous diver in the world. In 1943 he was a naval officer without a ship as all France's vessels had been interned by the Germans. He spent his time trying to make amateur movies with his skin-diving friends. Filming was continually interrupted because both cameraman and 'actor' insisted on returning to the surface to breathe. As his friends glided gracefully up and down, Cousteau envisaged a self-contained diving device unencumbered by air hoses and lead-soled shoes. Breathing air that continually flowed from a tank was ridiculously wasteful. What was needed was a valve that would deliver it only when required.

Cousteau's father-in-law was a director of Air Liquide, a gas engineering factory, and he introduced him to one of their brightest engineers, Émile Gagnan. Petrol was scarce in occupied France and Gagnan invented a valve that would deliver natural gas on demand to a car engine. The gas had to be stored in a huge bag lashed to the roof. Gagnan was able to modify the valve to one that would supply air at the same pressure as the surrounding water, but only when the diver inhaled. It was ingenious, simple and reliable. It was patented in both partners' names and within a decade the aqualung was being mass produced.

Neither Gagnan nor Cousteau was aware that an almost identical demand valve had been patented over a century earlier. Their compatriot Théodore Guillaumet built and demonstrated it by diving down to 52 ft (16 m) and staying submerged for twenty-five minutes. It supplied the diver with 'fresh air, the pressure of which corresponds to the pressure his chest is exposed to at the respective depth'. Members of the Académie des Sciences were impressed, so why didn't it become an essential piece of diving equipment? Well, Guillaumet didn't have Cousteau's entrepreneurial zeal and gift for self-promotion. Gagnan gave Cousteau the means to take the whole world underwater and, as often happens, the promoter gains the fame and the inventor is forgotten.

ON A WING AND A PRAYER

'Wilt thou now wing thy distant flight?'
– Emperor Hadrian

The first time I dived with an aqualung was a revelation. No longer shackled by gravity, I was weightless and at one with the flow, not swimming, but flying. We have long envied birds their effortless flight and even bumblebees buzz around despite being as aerodynamic as a Womble. The first aspiring aviators had taken only a cursory look at birds in action or they would have noticed that big birds spend much of their time soaring, not flapping. Yet the only ideas they embraced were feathers and flapping wings.

Unlike humans, birds lighten the payload by having hollow bones and stick-thin legs, while boosting their power-to-weight ratio with large chest muscles secured to a prominent keel on the breastbone. J.B.S. ('Jack') Haldane, a mathematician as well as a pre-eminent biologist, calculated that for a man to generate similar power, weight for weight, his breast would have to protrude 4 ft (1.2 m) to accommodate enough muscle to work his wings. So flapping was never going to succeed.

Pioneers were encouraged by angels' wings – although I suspect they are merely ceremonial. Thanks to Ovid it was well known that Daedalus had 'fastened feathers together with twine and wax . . . thus arranged, he bent them in a gentle curve, so that they looked like real bird's wings'. The flight went well until his son Icarus, intoxicated with aviation, ventured too close to the sun and 'The wax melted; his arms were bare . . . lacking wings they took no hold on the air.'

It was widely believed that feathers were not merely light in weight but also had an inherent tendency to rise, and it was *this* that held birds aloft. Another assumption was that birds 'swam' through the air by pushing it back and down. None of these well-known 'facts' was true.

Nevertheless, there were 'birdmen' aplenty willing to leap from tall buildings frantically flapping a pair of feathered wings. They discovered that flying was easy – if it is predominantly downwards. Most of them prematurely curtailed their inventing career. Among the earliest of these was King Lear's father, King Bladud, who believed he was omnipotent (as kings often do). His frailty was revealed when he jumped from a temple roof with feathers glued to his body and makeshift wings.

Some managed to glide a little and had softer landings. In 1742 a fifty-four-year-old French eccentric named Jean-François Boyvin de Bonnetot exited from a high window on an unscheduled flight across the River Seine. After gliding past the near bank he suddenly dropped and landed in a barge full of laundry. He only broke his arm.

The most remarkable flapper was John Damien, an alchemist to King James IV of Scotland. In 1507 he launched himself from the battlements of Stirling Castle strapped to flimsy struts covered with chicken feathers. Perhaps he hadn't noticed that chickens were not the greatest aviators in the bird kingdom. However, he managed to glide beyond the castle to land in a midden. He then remembered that chickens were earth-bound fowl and that was why he had homed in on a midden. In 2008 a Scottish historian found an old map that showed the nearest midden was half a mile away from the castle. Damien's other claim to fame came from being the patron of surgeons and physicians, and also of dentists, makers of dressings, apothecaries, druggists, 'well' nurses, physicists, haberdashers, hairdressers and confectioners. I suspect he had a good agent.

Kites

Damien's escapade resulted in a broken leg, but snapped bones were the least injuries that befell these over-optimistic 'tower jumpers'. None of them wore helmets or protective clothing. Indeed, if the illustrations are to be believed, several shunned *all* clothing. Perhaps it had something to do with improving their aerodynamics.

Damien may have been the first man to make a sustained free flight, but in Japan and China men had been lifted aloft on tethered kites for several millennia. The Emperor Kao Yang found great sport in strapping felons to bamboo kites and watching them plunge into the ground. When one stayed up long enough to be towed 1.5 miles (2.4 km) he was rewarded by being starved to death.

Kite flying could be useful. A fourteenth-century illustration depicts a kite-borne soldier dropping finned bombs into a castle. In the 1890s the American Samuel Cody, famous for his Wild West show which took England by storm, designed a stable box kite (a box-shaped frame open at both ends) that could carry a man in a basket high enough to view the deployment of enemy forces. It was adopted by the British Army.

However, tethered flight was not what the pioneers longed for, so they persevered with devices that emulated birds. In 1809 Jacob Degen, a Swiss watchmaker, built an 'ornithopter' (bird wing), although anything less like a bird's wing is difficult to imagine. His machine had two large conical flappers each bearing 3,500 strips of varnished paper. The pilot had to heave a horizontal bar as if he were a weightlifter. The paper strips separated on the upstroke and closed on the downstroke, thus trapping air in the cones. The greatest elevation Degen's ornithopter achieved was to hop around erratically on the ground and only when a balloon was attached.

In 1865 an American decided not to copy birds but to utilise them. He calculated that if an eagle could carry off a young lamb, ten eagles could lift a man. His idea was to hitch them to a circular frame with a man harnessed in the middle. Unfortunately, he had assumed that eagles were good team players whose sole mission in life was to hoist a man gently into the air. He soon found out that it was not:

A use for hot air

Inventors looked elsewhere for aerial propulsion. In the fourteenth century Albert of Saxony deduced that since fire was lighter than air, a craft filled with fire would float on the surface at the top of the atmosphere. For manned flight a fiery craft had certain drawbacks, but Albert baulked at further experimentation, not because of a surfeit of flames and smoke, but because of God's possible anger at a human rising above his station.

Nonetheless, fire *was* the solution. Heated air expands, reducing in density, and if channelled into an airtight envelope it floats upwards. In 1709 Bartolomeu de Gusmão amused the Portuguese court with a model hot-air balloon. As if by magic, it rose towards the ceiling and set fire to the curtains. The principle of the hot-air balloon was then forgotten until resurrected seventy-four years later by two French brothers. Joseph Montgolfier was a hybrid between a dilettante student and an absent-minded professor. He once checked out of an inn and forgot to take his horse. On another occasion it was his wife that was accidentally left behind. His brother Étienne was more down-to-earth – though not for long.

It was said that the idea of a hot-air balloon formed in Joseph's mind when he observed his wife's bloomers drying over the fireplace and billowing with warm air. He conducted several experiments with small, open-bottomed paper envelopes by slipping in scraps of burning paper. One of these trials was carried out in the sitting room, or the 'scorched room' as it became known. In 1783 Joseph constructed a much larger balloon with a light wooden frame covered with cloth panels lined with paper. The panels were buttoned together and since buttons need button*holes* the contraption was not ideal for confining air. To inflate, the open end of the envelope had to be held over a smoky fire, as Joseph believed (erroneously) that smoke was the 'lifting factor'. By the time the assistants were kippered the balloon was yearning to fly. The spectators were astonished to see it speed upwards until it vanished into the clouds. Ten minutes later, with all its eighteen thousand buttonholes whistling, it fell to the ground well over a mile (1.6 km) away. It was just the beginning of the year of the balloon.

High jinks

When the news reached Paris the academics were miffed that tradesmen (the Montgolfier brothers were papermakers) had invented a flying machine. A naturalist in the Natural History Museum who had 'effrontery to an eminent degree' sold tickets for a grand balloon launch even though he didn't have a balloon. He hired Jacques Charles, a bright young physicist who was mentored by Benjamin Franklin, American consul in Paris. Charles had never built a balloon before, but his first attempt possessed almost all the features of a modern gas balloon. He had no idea what had powered the Montgolfiers' balloon, but he did know the work of the English chemist Henry Cavendish.

Cavendish was brilliant and wealthy but eccentric in the extreme, having devoted his life to science at the expense of human relationships. All communication with his servants was via notes. They even had orders to keep out of his sight or they would be dismissed. One of his few friends said that he 'probably uttered fewer words in the course of his life than any man . . . not at all excepting the monks of la Trappe'. One of his many discoveries was the isolation of a new lighter-than-air gas that he named 'inflammable air', now known as hydrogen. Charles invented a method for producing inflammable air in large quantities. As Cavendish had predicted, his gas 'could lift a man into the heavens'.

The envelope was made of rubberised silk manufactured

by the Robert brothers to a secret recipe used for making condoms for the French royal court. The balloon's top half was covered with a large mesh 'hair net' to prevent it from bursting as it rose into thinner air; it also had a valve to vent gas. To avoid the traffic the semi-inflated balloon was carried through the streets of Paris on an ox cart in the dead of night. The sight of it must have persuaded a few tipsy stragglers to give up drinking. A taxi driver halted to let the cart pass and the driver knelt and took off his hat.

On the following afternoon a large crowd gathered in the square outside the Military College to see the balloon rise rapidly until it looked as small as an orange. It came to earth over 12 miles (19 km) away. Two hundred years later the Air France *Concorde* would crash at the same site. In 1873 the local peasants attacked the monster with pitchforks. A reporter recorded the balloon's death throes: 'The creature shaking and bounding, dodged the first blows. Finally, however, it received a mortal wound and collapsed with a long, sad sigh.'

Not to be outdone, Étienne Montgolfier built a 70-ft (21 m) high hot-air balloon and launched it from the grounds of the royal palace at Versailles. This time it was crewed by a lamb, a duck and a cockerel. They landed 2 miles (3.2 km) away. Only the cockerel seemed worse for wear, but it wasn't altitude sickness: the sheep had sat on him. The sheep was adopted by the Queen of France, Marie-Antoinette. Both the Montgolfiers and Charles had provided a woven wicker container beneath the envelope. Étienne had been winched up to several hundred feet in a tethered balloon, but who would be brave enough to surrender their security to the wind and a laundry basket?

The Montgolfiers' father pleaded with them not to be

the test pilots. Instead it would be Jean-François Pilâtre de Rozier and his friend the Marquis d'Arlandes. Pilâtre was a colourful character and a bit of a rake, but also energetic, charming and cool under pressure. He had invented a sort of gas mask for sewage workers and to test its efficacy he sat in a stinking sewer for a day and a half. The two men flew at low altitude, almost brushing the rooftops of Paris. All the time they were feeding the fire with straw to stay aloft and beating out burns on the balloon's cloth. When they spotted a promising landing place they damped the fire down and landed in a field 7 miles (11 km) from their starting point. The first manned flight had lasted twenty-five minutes and Pilâtre became the world's first test pilot.

Benjamin Franklin had been among the crowd at the launch and when the person beside him commented that ballooning was all very well, but what use was it? Franklin replied: 'What use is a newborn baby?'

Ten days later Jacques Charles and Marie-Noël Robert, one of the makers of the balloon's envelope, became the first to fly in a hydrogen balloon. A crowd of 400,000 saw them off at the Tuileries Gardens. Just before the launch Charles handed Étienne Montgolfier a model balloon filled with hydrogen and said: 'It is for you, sir, to show us the way to the skies.' Étienne released the balloon and as it sped away it was followed by Charles and Robert. They stayed airborne for two hours and travelled 27 miles (43 km). Both men were entranced. Robert called it: 'Perfect bliss . . . It's the sky for me! Such utter calm. Such immensity!' Charles said: 'Nothing will ever quite equal that moment. I felt we were flying away from the earth and all its troubles for ever.' He would soon change his mind.

On landing Robert disembarked and the loss of his weight caused the balloon to rush skywards with Charles still on board. Within ten minutes he had reached 10,000 ft (3,050 m), and entered a chill and lonely place where breathing was difficult and his ears were being stabbed by pressure changes. Slowly he released gas from the balloon and after what felt like an eternity he was safely back on the ground – but he never flew again.

In Paris people talked of little else than the great balloons. Almost half the city's population had been at the Tuileries launch. Toy balloons flew off the shelves, sophisticates sipped Crème Aérostatique aperitifs, when not swaying to the new dance named after the village where Charles's balloon had landed.

Tethered balloons appeared at fêtes and fairgrounds. Vicenzo Lunardi popularised ballooning in England with his pleasure trips. On one occasion the actress Laticia Sage and the dashing George Biggins took the ride alone. A viewer from the ground with a spyglass had a clear view of the couple through the open mesh passenger cage. Leticia was on all fours with George oscillating vigorously behind. Her memoirs confirmed that George experimented while aloft 'holding a device with two balls and exposing it to a cloud we were passing'.

Balloons were a novelty and few thought that they were a serious means of transport until a Frenchman and an American argued their way across the English Channel in 1784. Jean-Pierre Blanchard was the prima donna of pioneer ballooning. He was sponsored by Dr John Jeffries on the understanding that Jeffries would accompany him on the trip, but Blanchard was determined to keep the glory for himself.

He claimed that his balloon couldn't carry the weight of two passengers. The astute Jeffries suggested that perhaps Blanchard should discard the weight belt he was hiding beneath his shirt. Blanchard retaliated by refusing to take Jeffries' scientific equipment. They departed from the White Cliffs of Dover in midwinter, but their rivalry did not abate. Each contrived to drop the other's national flag overboard. A more pressing problem was that the balloon was progressively losing height until the basket was almost surfing. They jettisoned everything they could: the ballast, the steering paddle and their food. In desperation they even peed over the side and their winter clothes had to go as well. They arrived in France each wearing only a cork life jacket and breeches. Sadly no photographer captured this heart-warming moment.

In 1809 Blanchard died while trying to parachute from his balloon. His spunky widow took to ballooning and clocked up seventy flights. She became an aerial performer at fêtes. Her act was to be winched up sitting on a swing beneath a balloon while wearing feathers and a white hat, then to set off a firework display. It is worth mentioning that the balloon was full of 'inflammable air' and at a show over the Tivoli Gardens in Paris the inevitable happened, thus providing a spectacular end to her career.

Air mail

The event that proved balloons could be useful occurred in 1870 when Prussian troops surrounded Paris. Bismarck was sure that the isolation would soon bring Parisians to their knees, but the city's balloonists had a different idea. The first

balloon to leave the beleaguered capital landed 60 miles
(97 km) away and carried 275 lb (125 kg) of mail. It was the
first airmail postal service. The temporarily disused railway
stations were converted into balloon-making factories and
anyone with a head for heights was trained as a pilot. During
the year-long siege sixty-four balloons transported 155
passengers and over 2.5 million letters out of the city. They
also carried out homing pigeons to bring back messages.
Microfilming, invented by Englishman John Dancer in the
1850s, was used for longer documents. A single pigeon could
carry forty thousand pages.

However, the balloons were not flying to a post office: they
were heading wherever the wind took them. As the poet
Shelley said: 'The art of navigating the air is in its . . . help-
less infancy; the aerial mariner still swims on bladders.'
Balloons lacked a sense of direction, but that was about to
change.

Steering a floating elephant

Balloonists tried sails, propellers, winglets, even oars, but the
balloon remained stubbornly married to the wind. Henri
Giffard, a Frenchman, showed that a motorised balloon would
work if it was the right shape. In 1852 he built a streamlined
balloon that was 144 ft (44 m) long. A puny three-horsepower
steam engine turned a 12-ft (3.7 m) propeller and was carried
on a suspended cradle beneath the balloon. Giffard guided
the craft by means of a large, sail-like rudder. Wearing a frock
coat and top hat and standing on the cradle, Giffard launched
from the roof of the Paris Hippodrome. He made turns to
the left and right and then, with difficulty, a complete circle.

He was travelling at only 5 mph (8 kph) and admitted: 'Not for a minute did I dream of struggling against the wind; the power of the engine would not have permitted it.' Even so, he had flown 14 miles (23 km) on the first airship.

Dirigibles

In 1900 the Parisian public was amazed by the one-man dirigible balloon built by a brilliant mechanic, Alberto Santos-Dumont. He came to Paris from Brazil to study engineering, but was soon diverted by ballooning. Such was his passion for aviation that his dining table and chairs were suspended from the ceiling. Santos-Dumont's charm, daring and panache made him an instant celebrity. Although he was diminutive he was conspicuous, buzzing around the city dodging chimney stacks or alighting on a boulevard and hitching his dirigible to a tree or a convenient balcony and popping into the nearest café to chat with friends. He won a major prize by being the first to fly from the field of the Paris Aero Club at St Cloud to the Eiffel Tower and back. And most of all he demonstrated that a well-designed dirigible was a safe and exciting way to travel.

Piloting the craft kept the aeronaut fully occupied and Santos-Dumont complained to a friend that while aloft he was unable to get his watch out of his pocket. The generous friend responded by inventing the wristwatch for him. I suspect that timepiece might be quite valuable today – his benefactor was Cartier.

With big engines larger dirigibles should have carried larger loads. Unfortunately, the envelope drooped where the heavy engines were slung. The solution was not to have a single soft envelope, but several large, hydrogen-filled bags accommodated in compartments in a metal shell. This also had the advantage that a rigid metal skin created far less drag than a textile balloon. The rigid airship was the brainwave of a German, Count Ferdinand Zeppelin. When acting as a military observer during the American Civil War, he was taken up in a reconnaissance balloon by American aeronaut John Steiner. Soon afterwards Zeppelin germinated the idea of an entirely new type of airship and he patented the design in 1895.

The airships were built in a huge floating shed on Lake Constance. They were enormous: 420 ft (128 m) long. The early attempts were unsuccessful because the craft were heavy, underpowered and difficult to control. Zeppelin's brilliant designer, David Schwarz, decided that the framework of the fourth airship would be duralumin alloy, which was strong and only a third of the weight of mild steel. The metal struts had to be lagged with cotton to prevent static, as sparks were the last thing you wanted in a balloon heaving with hydrogen. The airship carried twelve people 240 miles (386 km) at an altitude of 2,600 ft (790 m) and became an emblem of German technology. When it was destroyed on the ground in a thunderstorm, a patriotic public donated the money to build another.

For five years before the outbreak of the First World War, thirty-five thousand paying passengers had flown safely in Zeppelins. But now the entire production of the factory went to the military. As early as 1908 Kaiser Wilhelm II had banned Zeppelin from selling airships to foreign countries. Airships replaced captive balloons for aerial reconnaissance and flew bombing raids over southern England. Their success can be attributed to Britain's pitiful defences against air attack. At the beginning of hostilities there were no anti-aircraft guns, searchlights or fighter aeroplanes that could reach the altitudes at which the Zeppelins flew.

After the war Zeppelin's company modified its bomber to take passengers. The daily service between London and Paris initiated the first-ever scheduled passenger flights. The airships could take very few passengers, so the cost of a ticket was exorbitant. By the 1930s airships were luxurious touring hotels. In 1909 no less a person than Wilbur Wright had been sure that: 'No airship will ever fly from New York to Paris. That seems to me impossible.' Within a decade a British airship crossed the Atlantic.

Zeppelins were the largest and most luxurious airships. The *Graf Zeppelin* was 774 ft (236 m) long, with the passenger quarters suspended beneath. It went around the world in only twenty-one days. Over its lifetime it carried thirteen thousand passengers and flew more than a million miles (1.6 million km). For the rich, the era of air travel in wonderful airships was born. But it was stillborn.

The *Hindenburg* was even bigger than the *Graf Zeppelin* and could accommodate seventy-five passengers. In 1936 as it approached its mooring mast in Lakeside, New York, a tiny flame appeared from the top of the airship. Within seconds the entire craft was devoured by fire. 'Inflammable gas' certainly lived up to its name. Newsreel footage of the disaster was viewed worldwide and sounded the death knell for airships as passenger liners.

Courageous and usually foolhardy pioneers were now trying to elevate heavier-than-air machines into the sky. Without a strand of aerodynamic nous, they assembled innumerable flimsy-winged boxes held together with a cat's cradle of string. They should have been better informed because after centuries of others' futile flapping Sir George Cayley, 'the father of aviation', had explained almost everything in the 1790s. He designed a manned fixed-wing monoplane glider and supplied the instructions for flying it. This had many of the features of modern aeroplanes. It had a tailplane, a large rudder, propellers, and an undercarriage with spoked wheels. It was so easy to steer that Cayley's pilot was his coachman, although he resigned, grumbling that he was paid to drive, not fly.

Cayley's understanding of thrust, drag and lift came from experimentation. He made several wing pieces with different cross-sections. A pivoted arm like a see-saw had a wing sample on one end and a counterweight on the other. It was set whirling and the downward deflection of the weight measured the lift generated by that sample wing. By also measuring the amount of lift generated at different angles of attack, he explained why an aircraft stalls. Cayley invented the science of aerodynamics and gave prospective aeronauts the ideal wing shape to lift an aeroplane. He was convinced that powered flight would come as soon as engines were sufficiently light and powerful.

The wind beneath the wing

Still, most of the contraptions given the title 'aeroplane' rarely left the ground, which was fortunate because when they did, the pilot hadn't the slightest idea how to control them. Otto Lilienthal, a German, was one of the few people who understood that control was vital and could be learned, but not in an aeroplane: on a hang-glider.

Leonardo da Vinci had designed a hang-glider. Almost five hundred years later enthusiasts built his machine and Britain's leading hang-glider pilot Judy Leden proved that it could glide, but Lilienthal is credited with making the first controlled flights, in the 1890s. The wings of his glider were shirt cotton stretched across elegantly curved willow struts. The wingspan was 23 ft (7 m) and there was a large tail fin for stability. Lilienthal controlled the glider's flight by swinging his legs to shift its centre of gravity. He had to be a fast learner, for his life depended on it: 'After a few leaps one gradually begins to feel one is the master of the situation . . . the aviator travels over deep chasms and soars far . . . without the slightest danger.'

It *was* dangerous, of course, so he added a stout willow hoop on the 'nose' to take some of the impact out of a crash. Within days the glider stalled and nose-dived from 65 ft (20 m). The hoop was thrust into the ground and smashed to pieces, but Lilienthal got away with just a sprained wrist.

He built eighteen gliders and strived to improve their performance – increasing the wingspan to 30 ft (9 m) and adjusting the configuration of the wings to gain more control. He reached an altitude of 820 ft (250 m) under controlled flight and completed 180-degree turns. In only six years he made almost 2,500 flights, until in 1896 he made one too many. A sudden gust stopped his glider as if it had hit buffers, and it plummeted to the ground. Lilienthal's last words were: 'Small sacrifices have to be made.'

More than anyone he combined an understanding of aero-dynamics with consummate skill as a pilot. He increased our knowledge of control methods and was experimenting with control surfaces such as flaps. There is also a legacy of stunning photographs in which he and his graceful machine are as one, soaring in the sky he conquered.

Lilienthal and others had shown that a single wing had substantial lift, yet pioneers building powered aircraft added to their weight with additional wings. Three and four sets of wings were commonplace and Count D'Ecquevilley's effort had seven sets of wings stacked one above the other like empty bookshelves. They managed to lift the Frenchman's machine 7 in (18 cm) from the ground. Some inventors ignored the necessity of reducing the load to an absolute minimum. Joseph Kaufmann, an engineer from Glasgow, based his design on the bumblebee, but his bee was burdened with a steam engine weighing 2.46 tons (2.5 tonnes), not including the weight of the fuel or the 90-lb (41 kg) stabiliser. It was the heaviest bee known to man – until he decided to upgrade to an engine three times heavier. The only way it was going to get off the ground was by crane. Its mechanical wings flapped so fast that they snapped. Most powered flapping machines vibrated themselves into a pile of junk.

The Wright stuff

Lilienthal's concentration on perfecting the controls before designing the aircraft inspired two self-taught mechanics from Ohio, who manufactured the very best bicycles, to study the mechanism of flight. Kill Devil Hills is a stretch of bleak sand dunes just behind the back of beyond on the coast of North Carolina. Its saving grace was that the chill wind never stopped blowing. That's what attracted Orville and Wilbur Wright. Having digested the theoretical basis of flight and been advised by a famous American aeronaut named Octave

Chanute, they began a series of experiments on how to control a machine in the air.

There are three main types of instability: pitching up and down, yawing to left or right and rolling from side to side. The Wright brothers' method was to test-fly kites, essentially unmanned planes, with wings of different shapes and sizes to determine which were the most stable. They tested around two hundred wing types for lift and drag. Some findings took them by surprise. Their designs had an aerofoil extending in front of the kite's nose and they discovered that it was more stable if flown backwards. Clearly tail fins were important.

Many pioneer aviators thought that an aeroplane should turn like an automobile on a flat road, but the Wrights realised that it was more akin to rounding a corner on a bicycle where the bike leans into the turn. This became known as 'banking' and during this manoeuvre the aircraft could become unstable. Fortunately, Wilbur had observed that when buzzards were rolled over by a gust of wind, they righted themselves by twisting their wing tips. He therefore built a mechanism to bend the trailing edge of the aeroplane's wing downwards. In addition to 'warping' the wing in this way they found they had to use the tail rudder at the same time to prevent the plane from rolling or swinging. Slowly and systematically they were inventing the control mechanisms that the pilot of a powered aircraft would have to master. They eventually graduated to flights in an unpowered glider. Wilbur thought of their aircraft as 'A fractious horse . . . and if you wish to learn, you must mount a machine and become acquainted with its tricks.'

There were doubters such as Lord Kelvin, who said: 'Heavier-than-air machines are impossible.' The novelist

H.G.Wells was more optimistic and believed that an aeroplane would probably fly before 1950. In 1901 even Wilbur Wright predicted that 'man will not fly for fifty years'. In December 1903 the Wrights' fragile *Flyer I* was lifted on to its trolley, which ran along rails to help the plane pick up speed for take-off. The track was laid on level ground, so the plane would have to ascend under its own power and its petrol engine was only eight horsepower.

Wilbur won the toss and would pilot the craft. But it didn't go well, as a witness wrote: 'A hurricane sprung up . . . everyone grabbed their hat and clung for dear life onto the rail. . . . Suddenly when the whole machine was shaking and straining on its anchor . . . the huge structure bounded across . . . at a rate increasing up to forty miles per hour [64 kph].' Despite its impetuous sprint it was airborne for only a few seconds and 'in a moment it lay . . . like a wounded bird with torn plumage and broken wings'.

It took several days to repair *Flyer I* before it returned to the sands. The winter wind was bitter and the plane would be taking off into a 27-mph (43 kph) gale. After a few false starts it stayed aloft for a second short of a minute and made a controlled landing 852 ft (260 m) away. Orville had flown into the future.

Unfortunately, the future for the Wright brothers was not what it should have been. Despite the photograph of their historic flight, it was dismissed in the media. One newspaper used the heading 'Flyers or Liars?' In 1905 they built *Flyer III*, which is regarded as the world's first practical powered aeroplane. It flew for over half an hour and covered 20 miles (32 km) at an altitude of 60 ft (18 m) and circled the Statue of Liberty, but it didn't sell in great numbers. Soon other inventors were taking out patents for new developments and the Wrights were left behind. Eight months *before* they flew for the first time, a New Zealander named Richard Pearse flew 500 ft (150 m) in his powered aeroplane, but since he guided it into a gorse thicket, it is not considered to be the first-ever *controlled* flight.

No land is an island

The *Daily Mail* offered large sums for the first crossings of the English Channel and the Atlantic Ocean. Several aeronauts ditched in the Channel and it fell to the Frenchman Louis Blériot in 1909 to zip across it, in thirty-six minutes. In consequence the sales of Blériot's aeroplane soared and Britons learned they no longer lived on an island. Only sixteen years after the Wright brothers' first flight two Englishmen, Captain John Alcock and Lieutenant Arthur Brown, made the first non-stop crossing of the North Atlantic in their Vickers Vimy biplane. They flew in an open cockpit for sixteen and a half hours through fog, sleet and snow. The instruments froze and bits of machinery fell off before they landed nose-down in an Irish bog.

SAFETY FIRST

'If a little knowledge is a dangerous thing, where is
the man who has so much knowledge as to be out of danger?'

– Thomas Huxley

Gravity is remorseless and the early aeroplanes were prone to descend prematurely. They were as robust as matchstick models and offered no protection for the pilot, who had no means of reaching the ground safely. The idea of floating gently down to the ground is an ancient one. Legend has it that four thousand years ago the Chinese emperor Shun was trapped in a blazing granary ignited by his father. The ingenious Shun joined straw hats together and parachuted to safety. How he attached the hats is unknown but one savours the image of him hastily stitching as the flames licked around his sewing basket. Another emperor, Shi Huang, was in the habit of jumping off the Great Wall clutching an umbrella whenever the fancy took him.

Get me out of here!

Leonardo da Vinci asserted: 'If a man have a tent of caulked [sealed at the seams] linen . . . he will be able to let himself fall from any height without danger to himself', but as usual he sketched his vision but didn't test it. Five years earlier an unknown Italian engineer drew in his notebook a conical cloth canopy with cross-struts at the mouth to keep it open and to provide handholds for the parachutist, who is suspended by straps fixed to his belt. In the late seventeenth century the King of Siam had an acrobat who amused him by leaping from a height aided by two umbrellas and landing on the ground, or in a tree, or in the river.

The first inventor to demonstrate what he called a 'parachute' (French for 'protection against falling') was Louis-Sébastien Lenormand, who in 1783 jumped from the observatory at Montpellier in the south of France. He had tested his device beforehand and given a variety of animals the privilege of going first. One poor sheep was dropped six times.

In 1797 André-Jacques Garnerin, 'a small peppery man' who had been a balloon inspector in the French army, took his life in his hands by making the first drop from a balloon. His ribbed canvas parachute was 30 ft (9 m) in diameter and accordion-pleated like a modern parasol. Garnerin stood in a basket attached to the parachute by a pole and at an altitude of 3,000 ft (910 m) he detached from the balloon. With the loss of weight the balloon shot upwards, giving the impression that Garnerin was falling much faster than he was. Adding to the drama, the balloon exploded. The watching crowd gasped and women swooned.

It was not an easy ride for Garnerin as the basket lurched violently and threatened to toss him out, and although he landed safely he was severely airsick. An observer called it: 'One of the greatest acts of heroism in human history.' Braver still was Garnerin's willingness to take several other jumps, including the first parachute drop in England when he jumped from 8,000 ft (2,440 m) over London and landed in the grounds of a smallpox hospital. It was a lucrative way to get airsick. A couple of years later Jeanne Labrousse became the first female parachutist. She fell for Garnerin and they married. Their niece Elisa also became a professional parachutist.

An Irish watercolour painter named Robert Cocking had witnessed one of Garnerin's displays and pondered how to eliminate the jerking and jolting that the Frenchman had suffered. His solution was an upside-down umbrella. Experiments with small models attached to tiny hydrogen balloons convinced him that his idea was sound. The full-sized parachute was 10 ft (3 m) high and 107 ft (33 m) in circumference. Three metal hoops supported an inverted cone of canvas braced with wooden spars. It weighed 223 lb (101 kg), not including Cocking's portly frame. To lift such a load he hired Charles Green's enormous balloon, the *Nassau*. At 5,000 ft (1,524 m) Cocking reassured Green with the words: 'I never felt, more comfortable or more delighted in my life . . . Well, I think I shall leave.' He pulled the release cord and fell away. He had taken the precaution of filling his trousers with inflated pig's bladders to cushion the impact of landing – they would not be enough. About 300 ft (91 m) from the ground his parachute collapsed. His corpse was carried to a nearby inn, where the entrepreneurial landlord charged sixpence to view it.

French astronomer Joseph Lalande explained the erratic misbehaviour of Garnerin's parachute. As it dropped, air accumulated in the canopy, increasing the pressure. The only way it could escape was by spilling out over the lip of the canopy. Spilling air on one side caused the 'chute to swing, so that air escaped on the opposite side. This set up an erratic pendulum motion. Lalande suggested there should be a hole in the top of the canopy to allow the air to escape smoothly. You might think that moths were a parachutist's greatest fear, but Lalande's big hole stabilised the parachutes of the future.

No matter how many parachutists landed slightly *below* ground level, there were always others waiting to have a go. Franz Reichelt, an Austrian tailor working in Paris, designed an overcoat that converted into a parachute. It looked more like a theatre curtain that had fallen on Reichelt and enveloped him in its folds. He explained to the authorities that he wished to test his invention by dropping a dummy from the lower platform of the Eiffel Tower. Permission was given, but at the last minute he strapped on his overcoat and jumped. There were no grey areas in pioneer parachuting: only wet, red ones. If Reichelt had departed from the top of the tower his parachute would have had sufficient time to open properly and he would have been the toast of Paris. As it was, he was just toast.

The first intentional leap from an aeroplane was taken by US Army Captain Albert Berry in 1912. His parachute was on a 'static line' that drags the parachute out *after* the pilot has leapt from the plane. When his parachute opened he was surrounded by a cascade of newspaper pages which had been interleaved with the folds of the canopy to keep them separate. Later that same year F.R. Law demonstrated the Steven's Life Pack parachute, which provided a ripcord to enable the pilot to open his parachute while falling.

During the First World War pilots of the Royal Flying Corps were not supplied with parachutes as the 'top brass' thought that airmen might abandon their plane prematurely instead of trying to coax it home. After all, there was a greater shortage of planes than of pilots. In the RFC's defence it's only fair to say that the 'Guardian Angel' parachutes that were available were so heavy they significantly reduced the aircraft's performance. Wesley May, an American flyer, was the first to attempt mid-air refuelling by leaping from the wing of one plane to the wing of another, with a can of petrol strapped to his back. The first successful refuelling involving two aircraft connecting doggy fashion was by US Army Service pilots while making a world endurance record in 1923.

The modern parachute was perfected by Leslie Irvine and Floyd Smith, civilian engineers working for the US Army. There was concern that at altitude the falling, spinning pilot would be unable to use his hands to pull the ripcord. Irvine proved that this was untrue by leaping from a plane. He had no difficulty operating the ripcord, but on landing he ripped his ankle tendons.

Flight Sergeant Nicholas Alkemade of the Royal Air Force had a gentler landing. He was the tail gunner in a Lancaster bomber who had to abandon his blazing aircraft at 18,000 ft (5,500 m) without a parachute. His fall was broken by a pine tree and then a snowdrift and he was found unhurt sitting in the snow smoking a cigarette. Even the cash-strapped Air Ministry didn't recommend this mode of escape to other pilots.

As aircraft flew ever faster it was no longer possible for pilots to climb out of the cockpit: they had to be forcibly ejected. In 1940 engineers in Sweden, Germany and Britain were working on ejector seats. The English firm Martin Baker set up a test rig on which weighted dummies were shot up a vertical tower to see how much explosive was required. They advertised for a guinea pig: 'Wanted – ejection seat tester. Involves a small amount of travel.' Bernard Lynch, a fitter at the factory, volunteered to be the first person to sit on the 'bang seat' and ride the tower.

The results indicated that a force of twenty-five G (twenty-five times the earth's gravity) applied 'gradually' over a tenth of a second would blast a pilot at a velocity of 59 feet per second (18 m/s) and clear of the plane's razor-sharp tailplane, without concertinaing his spine.

The original design required the pilot to pull a heavy canvas sheet over his head to prevent violent flexing of his neck. This action automatically set off a small explosion to remove the cockpit's canopy, and a much larger one to rocket the pilot and his seat aloft. A small drogue parachute stabilised the seat before pilot and seat parted company and the large parachute deployed. English test pilot Bryan Greensted made the first ejection in flight in 1945 and the following year Bernard Lynch ejected at 8,000 ft (2,440 m) from a Meteor fighter travelling at 320 mph (515 kph). Subsequently, Squadron Leader J.S. Fifield demonstrated a safe ejection from a racing fighter at ground level.

Belt up

Pilots who didn't bail out but attempted to land a disabled aircraft often sustained substantial injuries. Colonel John Stapp, a flight surgeon in the United States Air Force, was determined to see if better restraints might reduce their severity. He wasn't a man who did anything by halves, so in

1954 he tested his own tolerance to sudden deceleration in a rocket-propelled sled. It accelerated to 632 mph (1,020 kph) in five seconds and stopped in seconds – as abruptly as hitting a wall. Stapp suffered total blackout (seeing black when his eyes were open) and his eyeballs ballooned out of their sockets because when his restrained head halted, his eyes kept going. All his injuries and discomforts were temporary and he perfected the best possible harness to protect pilots.

He then turned his attention to motorists, who at that time were loose lumps of soft tissue rattling around in a tin box. In the year that Stapp sustained no permanent injury after coming to an abrupt halt from over 600 mph (900 kph), nearly forty thousand American motorists died from collisions at 25 mph (40 kph) or less. Clearly drivers needed to be belted in.

Seat belts were not a new idea. In 1885 New Yorker Edward Claghorn designed a waist belt to secure painters, window cleaners and firemen who on occasion dangled on the outside of buildings in a seat that could be hauled up and lowered. Claghorn's belt secured the workman to the seat and to the line from which it was suspended. In 1903 retired English civil servant William Dickson produced a safety strap for hansom cabs, to prevent the passengers lurching forward against the cabin or being flung into the road when there was an accident. It is astonishing that fifty years later, with cars travelling at far higher speeds than in the early 1900s, motorists were dead set against seat belts. As a result some of them ended up dead. American car manufacturers detested being told what to do by boffins who had carried out thousands of crash experiments and proved that seat belts made accidents more survivable. Volvo was the exception. It examined twenty-eight thousand accidents and found that unrestrained drivers had been killed even in collisions involving vehicles travelling as slow as 12 mph (19 kph), whereas for those wearing a harness there were no deaths in crashes at below 60 mph (97 kph).

The anatomical examination of the injuries sustained by motorists killed in crashes revealed that a lap belt was insufficient. The combination of a lap belt and a shoulder strap ensured that the pressure load was taken by the pelvis and the shoulder, both strong parts of the body. In 1959 Volvo technician Nils Bohlin designed the lap belt and the transverse shoulder strap we use today.

Many countries made the wearing of seat belts compulsory but that didn't always reduce the death rate of motorists. The explanation appears to be that a belted driver feels more secure and therefore takes more risks. Straps can restrain a man's body but not his testosterone.

Many American drivers refused to wear seat belts. The solution was to make the safety measure activate automatically when an accident occurs. In 1968 Allen Breed invented a device with sensors that detected a collision and triggered the instant deployment of an air bag. Breed's first model was so explosive that it packed quite a punch, so in 1991 he incorporated vents to allow a slight deflation to lessen the impact. It is still quite violent, so shunts are therefore best avoided by those with a long, delicate nose.

Do as you're told

Discipline on the roads was based on tradition. The tendency was to drive in the middle of the road and in Britain to move to the left when passing oncoming traffic. This was said to be because right-handed coachmen could readily defend themselves if the other driver became belligerent. In continental Europe the postilion, who guided the horses, sat on a horse at the left side of the team and preferred to pass oncoming traffic with his coach on the right-hand side of the road.

Traffic congestion has a long history; Julius Caesar banned chariots from the streets of Rome during the hours of daylight. Even in Victorian cities chaos ensued where streets crossed, with hordes of horse-drawn buses and carts vying for space with flocks of sheep and foolhardy pedestrians dashing between the vehicles.

The first road traffic signals were invented in 1868 by J.P. Knight, a British railway signalling engineer. They were installed close to the Houses of Parliament to enable the members to cross the road safely. A policeman operated a lever that raised or lowered the semaphore arms. A poster informed drivers that 'arms down' advised caution when crossing the junction, while arms out horizontally meant 'Stop!' for traffic on both sides of the crossing.

At night a gas lamp on top of the pole showed green or red instead of the semaphore arms. It worked well for a year, then the gas exploded, sending the pole skyward and the policeman to hospital. To the public this proved that traffic signals were more dangerous than traffic and it was fifty-eight years before traffic lights replaced policemen at London's road junctions.

The advent of cars and inept drivers added to the confusion on the roads. There was no driving test, although driving licences were issued by the Steam Boiler Association. The increasing traffic in American cities led to more accidents and Garrett Morgan witnessed a crash in which a small child was badly injured. He decided to do something to prevent accidents. Morgan, the poorly educated son of former slaves, made a living mending sewing machines. But in 1923 he invented the first traffic signals to be used in the United States. His traffic controller was first installed in Cleveland, Ohio, in 1914. It was a hand-operated semaphore system very similar to Knight's, but on his machine the words 'Stop' and 'Go' were painted boldly on the arms and if all three arms were vertical and showed 'Stop', that halted *all* the traffic approaching the intersection so that pedestrians could cross. At night there were also red/green lights. Shortly afterwards the city synchronised the signalling at groups of intersections to smooth traffic flow. Traffic lights became adopted throughout the world, even in Venice where two canals crossed.

In 1918 New York introduced the three coloured lights in a vertical box familiar to us today, but the signal still needed an operator standing in a high box in the middle of the intersection. Wolverhampton in England became the first town to have automatic traffic lights, which worked on a time-interval basis. It stubbornly stuck to a red/green system when the rest of Britain adopted red/amber/green signals. Eventually the Ministry of Transport forced Wolverhampton into line. There was universal agreement on red for 'stop' and green for 'go', although during Mao's cultural revolution China decided that red was better suited for 'go', so the traffic lights were changed – but only some of them.

There was also a need to improve road markings and no one contributed more than an English road mender named

Percy Shaw. He found driving along winding, unlit country roads at night was dangerous. Then one dark night he saw two bright green 'lights' ahead. They were the eyes of a cat. Like most nocturnal hunters, cats have highly reflective retinas to enhance their night vision and they reflected his headlights. Shaw had not just seen a cat: he had seen an opportunity.

His reflective road studs, trademarked and marketed as Catseyes, had two shiny marbles close together facing the traffic. They were encased in a strong rubber shell partly embedded in the road surface and set either on the mid-line or at the edge of the road. No matter how dark the night, the driver would never be disorientated again with Catseyes to guide him. Ken Dodd the comedian suggested that had the cat been facing in the other direction, Shaw would have invented the pencil sharpener.

The ingenious part of the invention was the narrow trough in the front of the shell where rainwater collected and every time a car ran over the rubber stud it flushed out water to wash the reflectors. In 1946 the Ministry of Transport purchased millions of reflective studs for Britain's minor roads. Shaw claimed he received only a farthing (a quarter of a penny) for every stud. Nonetheless, it made him a millionaire, although, apart from buying a Rolls-Royce, he continued to live a frugal life, sharing gossip with his old friends over a pint of beer and fish and chips wrapped in paper.

The railway system needed no reflective studs as its vehicles were confined to tracks, but it had been accident-prone since its maiden journey, when Stephenson's *Rocket* ran down a leading politician, who died hours later. Engine drivers couldn't swerve to avoid an accident or brake to a sudden stop, so an external control system was needed. At first they used railway 'policemen' stationed every mile (1.6 km) along the track. They were supposed to stay awake and alert day and night so that they could, by means of hand signals, waving a lantern or running a flag up a pole, inform the oncoming engine driver whether or not the line ahead was clear. Trains were controlled by time intervals, a system that depended on all the trains adhering to the timetable and travelling at the same speed. To avoid collisions the 'policeman' would halt a train if one had passed him in the previous ten minutes. If there was a breakdown, he was supposed to sprint down the track to wake up the previous man a mile away and tell him to halt all traffic. But if he did have to halt another train, would he then have to dash to alert the guy farther down the line? In an emergency, a bonfire had to be lit on the track. A really successful fire also consumed the wooden sleepers.

This was clearly a barmy system. Fixed signals were needed at critical locations to tell the driver to proceed or halt. In the 1840s signals were brought in, but not all of them were clear. On the approach to Reading station there was a dangling ball. This indicated that it was safe to continue, but drivers were told that 'if the ball was *not visible*, the train must not pass it'. No wonder there were accidents.

In 1841 C.H. Gregory, an engineer for the London and Croydon Railway, invented the semaphore signal. The post had two arms, one for each direction. If the arms were horizontal they signalled 'Danger' and, if lowered to forty-five degrees, 'Caution'. The arms when upright were concealed in a slot in the post, meaning 'All clear'. Later it was simplified to just 'Clear' and 'Stop'.

Trackside men still had to change the signals and the points to direct a train on to the correct track. By 1856 the points and the corresponding signal were both changed by pulling a single lever. Later all the controls in an area were brought into a signal box from which the signaller had an overview of all the lines under his control. Gregory also devised a mechanism to prevent two signals giving clearance that would put trains on a collision course. Several railway companies introduced electric telegraph connections between operatives to improve communication.

Accidents should have been a thing of the past, were it not for human error. The inquiry into a head-on crash in Radstock, Somerset, revealed that the telegraph clerk was exhausted after a fifteen-hour shift and the operator with whom he was communicating didn't know how to read the telegraph's instructions. In 1865 a train derailed and plunged into a stream at Staplehurst in Kent. A track gang had been given eighty-five minutes to replace timbers on a bridge before the next train was due. But the gangmaster failed to check whether there were any irregular trains not on the timetable. The approaching boat train was one of these. The gang had set no warning signals and when the train arrived the bridge had been repaired, but the track had not yet been re-laid. Charles Dickens and his mistress, Ellen Ternan, survived the crash but the writer never recovered psychologically. Some believe this contributed to his premature death only a few years later, which left the ending of *Edwin Drood* a mystery. In Britain's worst-ever train disaster, at Gretna in Scotland in 1915, the signaller had forgotten where he had parked a train even though it was in full view from his signal box. The consequence was a three-train pile-up.

It took forty years for the railway companies to ditch the

patently ineffective time-interval system for keeping trains apart. The 1889 Regulation Act recommended three procedures: lock, block and brake. Locking involved ensuring that the signals agreed with the setting of the points; blocking meant allowing only one train to occupy a given stretch of track; and braking called for the installation of automatic brakes on all coaches. The first two measures were implemented promptly but the third was not completed until the 1940s.

For those in peril on the sea

Buoys with their lights and warning bells are the traffic signals of the sea and on the approach to a harbour the safe route is often lined with them. Floating lights have been around since Roman times. Galleys with fire beacons were put outside harbours when needed. It was not until 1731 that what was probably the first permanent floating light was anchored in the approaches to London. It was a ketch with a lantern that was raised at night. Subsequently, lightships were sited wherever there was insufficient rock to build a lighthouse.

The first land-based lighthouse was one of the wonders of the ancient world. The pharaoh Ptolemy wanted to create a great seaport at Alexandria and built large breakwaters. He also commissioned the Greek architect Sostratus to construct a great tower to mark the entrance. The Pharos of Alexandria was estimated to be 350 ft (107 m) tall. It was built of large stone blocks locked together vertically with molten lead ties, and it was topped with a large fire basket. The Arabian geographer Edrisi wrote: 'It burns night and day for the guidance of navigators and is visible at the distance of a day's sail.

During the night it shines like a star; by day you may distinguish its smoke.'

Its wall bore a fulsome tribute to Ptolemy the lighthouse builder, but Sostratus took the risk of engraving his name on one of the stones and covering it with a thin layer of mortar. He knew that over time the mortar would weather away to reveal the true identity of the architect. His pharos survived for 1,500 years until it was toppled by an earthquake. Diving archaeologists have found the remains of the tower in Alexandria's harbour.

The first modern lighthouse built off a coast was constructed over three hundred years ago on a perilous reef off Cornwall named Eddystone Rocks: the 'evil demon' that sank ships en route to Plymouth. Henry Winstanley, an architect who had been the understudy to the Surveyor General, Christopher Wren, drew plans for a lighthouse on the rocks. Building his tower was difficult as the men could work only on a receding tide on calm days. According to Winstanley, even in summer 'the weather at times would prove so bad yet for ten or fourteen days ye sea would be so raging round these rocks'. Even if they could land they might be able to work for only two hours or so. To row out to the worksite could take up to nine hours.

And they had more than just the weather to contend with. England was at war with France and a passing French privateer kidnapped Winstanley. To Winstanley's surprise, Louis XIV apologised and offered him a job. He declined and two weeks later he was back on Eddystone Rocks.

With the tower half-built he and his crew were marooned on the site for six weeks and were 'almost at the last extremity for want of provisions' before they were relieved. When they returned the next year some of the mortar at the stone base had worn away and they had to build a wall around it, as well as strengthening the wooden upper part of the tower with metal bands. In 1698 the sixty candles in a candelabra housed in the glazed lantern room were lit for the first time.

The tower withstood the waves for five years. In November 1703 Winstanley went out to it to oversee repairs. The wind rose to hurricane force and giant waves demolished the tower bit by bit. The lighthouse builder and his men were swept away. The storm also shook houses miles inland and at Winstanley's home a beautiful model of his lighthouse fell to the floor and was smashed to pieces.

The loss of the Eddystone lighthouse put ships in jeopardy again, so another wooden lighthouse was erected on the reef. On the very day that a shipment of gin arrived for the lighthouse keepers a fire broke out. For a while it was the brightest beacon on the sea. The only sober keeper tried to douse the fire in the lantern room. Gasping for air, he looked up and molten lead from the roof fell straight down his gullet. The post-mortem found a flat piece of lead as big as a man's palm in his stomach.

A third lighthouse was designed and built by twenty-three-year-old John Smeaton. From childhood he had been fascinated by mechanical devices. He invented a pump and tested it by emptying his father's precious garden pond, and he would go on to design the ultimate diving bell, which used air supplied by one of his pumps.

Smeaton's elegant tower was constructed entirely of stone, and tapered from a broad base to the top, which gave it a low centre of gravity and great stability. Some basal stones were dovetailed into the rock and every course of stones was secured to the course below with marble dowels. Smeaton made the template from which all later lighthouses would be built. His tower withstood the worst seas, but the rock beneath began to weaken, so his tower had to be dismantled and replaced. A young lad who was one of the dismantling team fell 70 ft (21 m) towards the jagged rocks, but at that instant a large wave rose over them and swept him into the sea. He swam back to the rocks having sustained only a few cuts and bruises. Smeaton's lighthouse was rebuilt on Plymouth Hoe as a monument to his vision.

The 'Lighthouse Stevensons' took on Smeaton's mantle and improved upon his innovations. Robert Stevenson, who founded the Edinburgh firm, had three sons, all of whom became engineers. Thomas, who was the father of the writer Robert Louis Stevenson, invented a superior type of lens.

The lighthouse buildings were perfect but their lights were feeble. Surely there was something better than candles. Oil lamps were brighter, but still most of the light was wasted, whereas it needed to be beamed out to sea. In 1757 Jonas Norberg, a Swede, designed a parabolic mirror which, if placed behind the lamp, focused the light into a beam. He claimed that the light could be seen 8 miles (13 km) away. He later invented a clockwork-driven mechanism that turned the light so that it emitted a bright flash at pre-set intervals. Mariners identified a lighthouse by the frequency of its flash, thus making it an aid to navigation.

In 1789 François Argand, a student of the famous French chemist Antoine Lavoisier, invented an oil lamp that gave a much brighter light. He was dining in a typically dimly lit room when a glass flask was broken. He picked up the neck of the flask and held it over the flame of an oil lamp. In his own words: 'Immediately it rose in brilliance.' He instantly saw the possibilities and designed a lamp with a glass funnel and a hollow wick so that the air could be drawn up inside the wick to feed the flame. Argand lamps banished the crepuscular gloom of living rooms until, over two centuries later, we discovered 'mood' lighting.

Large models of Argand's lamp with concave mirrors behind were used in lighthouses. Some lantern rooms housed thirty, multi-wicked Argand lamps. The heat was tremendous and the windows rapidly became obscured by soot. Overnight, lighthouse keepers were converted into window cleaners.

The next breakthrough was also made by a Frenchman, Augustin Fresnel. Every schoolboy knows that a magnifying glass can set paper alight. Fresnel produced wafer-thin prisms that refracted light into a beam. Their slight weight meant that they could be assembled on to large, concentric arrays of hundreds of lenses. These prisms greatly magnified the light and increased the range of the beam.

As a child I fell in love with a lighthouse when I climbed the hundreds of steps up to its lantern room. The light was encased in a wilderness of prisms like a crystal from another planet, able to incandesce and brush aside mere earthly darkness to explore my bedroom ceiling 2 miles (3.2 km) away. The huge lamp floated on a lake of mercury and the slightest nudge would have made it revolve.

Some arrays of lenses using Fresnel prisms weighed 20 tons (20.3 tonnes) and had an incandescent source emitting a light equivalent to well over a million candles to 'flash their splendour across the wastes of the night'. Even when the tower is out of sight over the curvature of the earth, it glows like a hidden fire.

Irrespective of lights and buoys, people still find themselves in the sea. Sailors are notorious for not being able to swim and pilots never think they are going to dunk into the ocean. They rely on an inflatable life jacket to keep them afloat until rescued.

Saving life at sea was a Victorian and Edwardian obsession, but in those days the solutions were not always practical. Perhaps the least useful device was a sort of wheel-less water cycle invented by Frenchman François Barathon. In the illustration a lady sits upright on a seat attached to a long, vertical pole, the upper half of which is a mast for a sail to keep her going, while the lower half holds a small horizontal propeller with another propeller behind her to augment the sail. She works one with pedals and the other with hand-cranks. I assume that the horizontal propeller is meant to thrust her upwards. If so, it's doing a great job as she is sitting well clear of the surface, with only her shins and feet beneath the water. There is no sign of any buoyancy, only metal pipes and cranks. Even the ill-fated passengers of the *Titanic* would not have complained if there hadn't been sufficient cycles for every passenger.

Gentleman inventor Henry Hallock appreciated the importance of buoyancy but he realised that the passengers didn't need personal buoyancy if their state room could float. He imagined round rooms, anchored into grooves on the deck, which would be released automatically if the ship sank. The survivors wouldn't have to suffer the privations of a crowded lifeboat, but instead would relax on a chaise longue while gently bobbing on the briny.

Other inventors appreciated the importance of buoyancy for the individual. One suggestion was for buoyant shoes

– which would be ideal for floating *upside down*. Carl Biebers invented a hollow walking stick that contained two pigs' bladders that could be extracted, inflated by mouth and tied together to make a translucent dumbbell to support the survivor – not the only dumbbell involved in this invention. Frenchman Léon Lejust went even more inflationary. His survivor is portrayed wearing an inflated shirt and trousers and is so buoyant that his top hat is not even splashed and his cigar still glows.

Cork is almost as buoyant as air and doesn't leak out or go soggy. In 1841 Napoleon Guerin of New York City patented a life preserver which was more like cork couture underwear. It was a snug tunic and if his illustration is our guide it was best worn with a jaunty beret. However, unlike many 'life preservers', it looks as if the wearer might float.

The cork-filled, ring-shaped lifebuoy, seen in coastal areas hanging with its coiled rope, was the brainchild of a Yorkshireman, Mr Cate. It is supposed to be thrown at someone in difficulties in the sea. The potential drownee ducks underneath the ring and pops up within the life belt so that someone onshore can haul him or her to safety. But what if you throw the lifebelt and it hits the already distressed swimmer and knocks him out?

The modern inflatable life jacket that the stewardess demonstrates before every flight was developed by several manufacturers. While serving on a rescue launch Englishman Edgar Pask was stunned to see the number of bodies wearing life jackets yet floating face downwards in the water. As a result of his observations designers came up with several different types of 'Mae West', so called after the voluptuous film star who looked as if she had something inflatable hidden about her chest. They were all supposed to force the wearer on to his back.

Pask ran a series of experiments to test their self-righting ability. The idea was for him to float face down and see if the jacket turned him over, but he found it impossible to stay inert and not help the jacket to do its job. So he was anaesthetised and, although he could still breathe with his face underwater, he was unconscious. The best-performing jacket was taken through its paces in the wave-making tank at Elstree Film Studios near London to ensure it worked even in choppy water and waves over 3 ft (1 m) high.

Where there's smoke . . .

Fortunately for Alexandria, it boasted not only a great lighthouse but also brilliant inventors. Ctesibius, the son of a barber, invented the first water pump designed to put out fires. It was not able to save Alexandria's library when it was torched, and the book describing all his inventions was lost. Hero of Alexandria improved the pump with a rocking arm. As the downstroke of the arm caused a piston to propel a jet of water out of an exit pipe, the uplift of the other arm sucked up water from an adjacent tank.

The growth of cities with narrow streets and wooden houses huddled together made them kindling awaiting a spark. So ancient Rome had the first firemen and their fire engines carried Hero's pump. They were manned by slaves who were said to be averse to danger and therefore they didn't arrive until after the conflagration had died down. One gang of criminal firemen, led by the appropriately named Crassus, rushed to the scene and offered to buy the property or they would let it burn to the ground.

Organised fire brigades vanished for a thousand years after the fall of the Roman Empire. Of the Great Fire of London in 1666, Samuel Pepys said: 'There was no manner of means to quench the fire.' At first the mayor of London was unimpressed by the blaze: 'A woman might piss it out,' he said. What a pity she didn't, because it went on to consume eighty per cent of the city. When it was rebuilt fire plugs were put into the water mains to provide a source of water for future firefighters. Nicholas Barbon introduced fire insurance in 1672. Barbon was actually christened 'If-Jesus-had-not-died-for-thee-thou-hadst-been-damned' by his eccentric, preacher father 'Praise-God'.

In the same year Dutch artist and engineer Jan van der Heide, who was the chief of Amsterdam's fire brigade, used leather hoses for the first time to douse fires and Englishman Richard Newshaw patented a powerful fire engine that could feed several fire hoses at the same time.

In Britain in the eighteenth century Acts of Parliament made it compulsory for fire engines to carry a ladder that would reach three storeys high. But what about the poor souls trapped above the third floor? Many inventors tackled the problem of rescuing people who were out of reach. As early as 1766 David Marie, a London watchmaker, proposed that several could be saved all at once in his wicker basket which was lowered from the roof by means of pulleys and chains. His system was not unlike the one used today to lower lifeboats from ships.

There were impractical schemes by the dozen. Benjamin Oppenheimer, an American, plumped for parachutes, and confidently predicted: 'A person may safely jump out of the window of a burning building from any height and land without injury.' He designed boots with elastic soles to absorb the impact of landing. His parachute was waxed cloth almost 5 ft (1.5 m) wide and attached to a frame of metal struts. A novel feature was that the struts were joined not to his body, but to a helmet fastened under the chin by a leather strap. It seems unlikely that Oppenheimer tried to get out of a window wearing this huge structure. If he had, the jolt of the jump would surely have wrenched off his helmet or perhaps his head. Thus relieved of so much weight, the parachute would have risen in the updraft from the fire below and drifted away to land in someone's garden and spoil their picnic.

If the building had stocked too few parachutes, two Englishmen had an alternative exit strategy. Ralph Jones and John Hodges invented a large reel with rope wound round it. The rope was firmly attached to the wall and the escapee was instructed to strap on the reel and jump. The rope would unwind on the way down, with the terrified faller desperately trying to apply the brake to slow his descent. Wouldn't it have been better to secure the heavy reel to the wall and tie the rope around the waist?

As recently as 1976 Arthur Pedrick, an English serial patenter, recommended huge reels on the roof of a skyscraper. If fire broke out, the reels would automatically roll out non-inflammable curtains down the sides of the building, snuffing out the fire from lack of oxygen. As the people inside would also need oxygen, Pedrick suggested that they rush to designated survival rooms. When the curtains fell, holes in them would halt alongside the window of the survival room.

Another inventor devised an even more elaborate way to

evacuate a theatre in case of fire. The weight of fleeing folk somehow caused entire galleries to swing out and be lowered to the ground. The pit area would be 'removed bodily by means of a rack and pinion mechanism'. The inventor was a brush manufacturer who would have made a killing selling brooms to sweep up the bodies of panicking patrons. I advise theatregoers to shun auditoria fitted with this mechanism. Safer by far was surely the 'fireproof' Iroquois Theater in Chicago, which had taken on board all the latest fire precautions. In 1903, a month after it opened, a faulty light bulb set fire to the scenery and the fire curtain jammed, so the building was razed to the ground.

Fire was nothing new to Chicago, for in 1871 the entire city had been destroyed by fire. It must have been contagious because Boston burst into flames the very next year, and after a decent interval Baltimore and San Francisco were also destroyed by fire; it was San Francisco's third experience of devastating city-wide flames.

City firemen needed all the protection they could get. In 1823 Charles Deane, inventor of the fire helmet mentioned earlier, patented a smoke hood supplied with fresh air through a hose to allow firemen to work in smoky conditions. But it was a commercial flop. Thirteen years later Colonel Gustave Paulin, commandant of the military fire brigade in Paris, saw eight of his men suffocated in a warehouse fire. It encouraged him to design a smoke-proof dress. This allowed firefighters to enter smouldering cellars 'in the merchants' district where they store alcoholic liquors, sulphur, resins and other such commodities'. His men could 'stay without danger for half an hour in heat of fifty degrees centigrade'. Paulin's invention was awarded a prize by the Académie des Sciences.

Around 1820 John Roberts, a 'common miner', invented a different sort of fire dress. It was a hood and tunic made from two layers of flannel, with glazed, tin-rimmed eye pieces. The layers were kept separate and away from his head and face by rods protruding from a helmet by as much as 3 in (7.6 cm). The flannel was thoroughly soaked with water before being put on.

In 1823 he demonstrated his dress in an airtight room at a Vitriol works in Whitehaven, north-west England. A copper cauldron in the room was filled with smouldering damp straw and sulphur. Roberts strode in and thirty minutes later the door was opened and observers rapidly retreated from the noxious fumes. Roberts seemed none the worse for his stint in an unbreathable atmosphere. To keep his flannel cool and damp he had strapped on his head a sponge which he squeezed from time to time. An observer commented that his stay in wet clothes would surely 'render him liable to colds'. Roberts's invention earned him a gold medal from the Society of Arts.

Modern firefighters have far better equipment than a flannel suit and they need it because they are always in danger, as the firemen of Arklow in Northern Ireland know well. In 1984 they were called to a blaze at their own fire station but arrived too late to save it. It was particularly galling because it was the second time that it had burnt down.

SEEING THE LIGHT

'Electricity is the soul of the universe'

– Nietzsche

Alchemists discovered that rubbing fur on amber enabled it to pick up small pieces of parchment, much as a magnet attracts iron filings. The friction of rubbing generated what came to be called static electricity, the force that makes your hair stand on end when brushed or when you get a shock from touching a car; and it's *your* static discharging, not the car's. William Gilbert, the physician to the first Queen Elizabeth, named this power to attract 'electrica', after '*electron*', the Greek word for amber.

Spare the rod and spoil the lightning

Mankind had always been aware of a more spectacular source of electricity, a gigantic spark that comes from the clouds. Lightning travels at 1,000 miles per second (1,600 km/s), creating a sonic boom we call thunder. It strikes at the tallest thing in the landscape. I have seen trees seared from top to bottom. Golfers may foolishly shelter beneath trees when thundery showers threaten and a flash of lightning livens up a dreary round. It generates temperatures five times higher than the surface of the sun but it passes through the golfer's body and discharges its energy into the ground. Providing it doesn't visit his heart or spinal cord en route the experience is not fatal. An American park ranger has been struck seven times and only suffered a temporary loss of toenails and eyebrows, burning hair and a slight roasting of his shoulder and leg.

Men are struck six times more than women even though they are not that much taller than females. I read recently of a fellow who bought an umbrella hat on eBay and the first time he wore it he was struck by lightning – twice. Arguably, anyone who dons an umbrella hat deserves all he gets. If you don't want to be a mobile lightning conductor, do not go out during an electrical storm carrying a golf club or wearing an under-wired bra.

The Roman philosopher Lucretius was on the right track when he described lightning as 'rarefied fire . . . composed of minute mobile particles to which absolutely nothing can bar the way'. But there is evidence that the ancients succeeded in taming lightning. Copper-coated masts on the gateways of ancient Egyptian temples were lightning conductors. An inscription on one temple states that the masts were 'arranged in pairs to cleave the thunderstorms in the heights of the heavens'. A Greek archaeologist has suggested that the spear-like metal rods on some Minoan temples served a similar purpose. The idea of coaxing down lightning to dissipate safely in the ground was lost until the eighteenth century.

American polymath Benjamin Franklin was one of seventeen children and left school at ten. His drive and intelligence enabled him to succeed in fields as varied as science, politics and diplomacy. With no scientific training he made significant observations on meteorology, sea currents and electricity.

Franklin's most famous experiment was to fly a kite during a thunderstorm to test whether lightning was a form of electricity. Voltaire had warned: 'There are some great lords whom one should approach with extreme caution: lightning is such a one.' Nevertheless, dozens of heroic paintings depict Franklin lashed by wind beneath a lowering sky, tempting a deadly bolt from the clouds. Curiously, he never wrote up his results and Tom Tucker, a distinguished science writer, has made a compelling case that Franklin's experiment was never carried out. In a letter to a friend, which was later published, Franklin asserted that it was an experiment 'which anyone may try'. Sadly, some of the scientists who tried were electrocuted.

As the French writer Balzac put it: 'Franklin invented the lightning rod, the hoax, and the republic.' His laboratory experiments with low levels of electricity showed that sparks were attracted to metal rods and the ones with pointed ends discharged electricity more efficiently than those with blunt tips. A rod fixed on top of a tall building and attached to a thick wire would direct the electric charge down into the ground.

Governments were convinced that lightning conductors would, in Franklin's words, 'secure us from that most sudden and terrible mischief'. In America lightning rods sprang up on roof tops, although the clergy condemned these attempts to avoid God's wrath. In France local authorities sued those who erected rods as they would *attract* lightning. In England a committee of the Royal Society concluded that pointed rods were better and should be erected on all tall buildings. One of the panel's members dissented and claimed that knobs were superior; he even persuaded 'mad' King George III. The heart of the matter was that Franklin had signed the American Declaration of Independence and was now one of the enemy. The king ordered that all the Yankee points must be replaced with patriotic knobs, but the government ignored him. In France fashionable and wary women attached lightning rods to their *chapeau*, with a conducting wire trailing on the floor.

Volta

Scientists were unable to carry out experiments on electricity until there was a portable source of it. But how could such a fluent substance be stored? The solution came from two very different Italians. Luigi Galvani was a conservative anatomist who believed he was uncovering God's works, whereas Alessandro Volta was a young physicist and lover of women who was certain that reason and rationality would supersede religion. They both puzzled over something seen when Galvani was dissecting a frog.

A room in his house was littered with dead frogs, many of them destined for the dinner table. When he touched a dissection with his scalpel and tweezers the dead frog twitched. He believed that he had stumbled across the life force. But Volta thought that the two different metals in Galvani's dissecting tools 'by themselves . . . excite and dislodge the electric fluid from its state of rest'. To test his idea Volta placed a silver coin on his tongue and a patch of tinfoil beneath the tongue, the two linked by a short copper wire. It immediately generated a sour taste which, he said, 'shows that the flow of electricity from one place to another is continuing without interruption'. The two metals were making an electric current.

155

He built a 'Voltaic pile', composed of plates of zinc and silver one on top of another in pairs separated from other pairs by cardboard discs soaked in brine. A copper wire joined all the plates together to make a battery. The larger the pile, the more electricity it produced; each stack had up to sixty pairs of plates. Volta had invented the first battery that could deliver a continuous, sustained supply of electrical current. However, had he visited the Baghdad museum he would have been surprised to see a small, two-thousand-year-old jar that contained an iron rod in a copper cylinder. Archaeologists believe it was a battery and experiments with replica jars containing vinegar or wine did indeed produce a small current.

Volta demonstrated his device to Napoleon, who was so impressed he commissioned a battery with 600 piles to be installed in the École Polytechnique in Paris. The brilliant young English chemist Humphry Davy built an array of 800 piles covering an area of 26 ft (8 m) by 13 ft (4 m): the biggest battery ever made. Within a week it was separating hydrogen and oxygen from water. Electrolysis had been invented and the composition of chemicals could now be analysed.

Volta was awarded the Royal Society's highest medal and Napoleon made him a count. The volt, the unit of electricity, was named after him. Not bad for someone who didn't talk until he was four and was thought to be retarded.

Meanwhile Galvani was hanging dead frogs on his garden railing when thunderstorms were due, to see if lightning would revive them, a trick only achieved by Baron Frankenstein. Galvani's nephew Giovanni was doing the next best thing: reactivating the corpses of executed felons on stage by shocking them. If he put his probes in the top of the spine and up the rectum the cadaver sat bolt upright and gave every impression of being about to leap off the table. Hence the verb 'galvanise' – to stimulate into action. It was suggested jokingly that Giovanni might revive convicts faster than they could be hung. The detached head of a guillotined murderer could only be coaxed into winking, but that was quite enough for most of the audience.

If electricity could pep up the dead, it must surely be a tonic for the living. The public became obsessed by electrical cures and, overnight, quacks became 'electrical therapists'. They had a cure for everything from rheumatism to 'wilting passion'. There were Galvanic socks, belts, corsets and hat bands. Getting wired up and shocked was claimed to rid the blood of 'diamagnetic poisons', thus invigorating the patient. The best-selling massagers were also invigorating when applied to the body's nether regions. One inventor used a generator to service thirty patients at a stroke.

There were numerous 'electric baths'; one 'intended to influence the action of the heart' – by stopping it perhaps. Later a 'therapeutic bath' provided 'high frequency electricity' plus X-rays and 'radium emanations'. The inventors warned that there might be danger – of the reflective canopy falling on the bather. Should the patient not survive the treatment, 'galvanoplasty' (electroplating) could cover the departed in a thin coating of copper before the internment. It was very popular in France.

Scientists dreamed of far better uses for electricity and one in particular powered a revolution. Michael Faraday, an ill-educated son of a blacksmith, had a love of facts and became Humphry Davy's assistant at the Royal Institution in London. He was the first to characterise the group of chemicals that made plastics and synthetic fibres possible, and went on to invent stainless steel. His experiments resulted in a machine that generated an electric current by magnetism. He invented the dynamo, the generator and the transformer, yet he greatly underestimated his achievements: 'It may be a weed instead of a fish that, after all my labour, I may at last pull up.' Faraday had transformed electricity from a novelty into the driving force of the modern world. The great physicist Sir William Bragg wrote: 'Prometheus, they say, brought fire to the service of mankind; electricity we owe to Faraday.'

Scientists could now harness electricity and many inventors attempted to improve artificial lighting. For millennia the world had been lit by candles and little sooty oil lamps. The early street lights had a single candle, replaced every day. When Brunel's navvies were excavating the 2-mile (3.2 km) long Box Tunnel, near Bath, they used 143,000 candles a day for two and a half years.

Candles indoors could be dangerous. Queen Caroline's physician stooped to lance her arm for blood letting when a candle set his wig on fire. The procedure was delayed until the Queen stopped laughing. By the mid-nineteenth century many houses had been fitted with gas. I once lived in a house with hissing, harsh gas lighting and the annoyingly fragile mantles that surrounded the flame. When gas street lights were introduced, ladies of the night complained it was ruining their business.

Let there be light

Far superior illumination had been invented long before. In 1807 Humphry Davy had harnessed the three thousand volts from his array of voltaic cells to demonstrate an arc lamp. It involved two carbon rods that touched briefly and were then withdrawn ever so slightly so that a continuous spark leapt between the rods. The white-hot tips of the rods gave a brilliant light for as long as the current continued.

Blackpool is a seaside resort renowned for its spectacular illuminations. At its very first light display in 1874 the crowds were astonished when night became day thanks to just eight arc lamps. Electricity was ideal for street lighting and the first place to be lit by it was the Place de la Concorde in Paris in 1841.

Arc lamps were far too bright and expensive for indoor use, but as early as 1801 Davy had demonstrated the first incandescent light by sending a current through a strip of platinum until it gave off light. Putting a warm glow in the parlour was, however, some way off. The principle was simple: if electricity was passed through a material with a resistance that made it heat up, it became incandescent – that is, it emitted light. Getting it to glow was the easy bit. Seventeen inventors got that far, but didn't persevere to develop a lamp for domestic use. The problem was that when a substance is heated up sufficiently to produce light, it burns away and there was no market for a ten-minute lamp. The obvious solution was to put it in a glass container and remove the oxygen that supported combustion, but there were no pumps capable of sucking *all* the air out of a vessel.

The breakthrough came with a young pharmacist in Newcastle upon Tyne named Joseph Swan. In his spare time he was an amateur inventor and while still in his teens he patented chemically treated paper for printing black and white photographs.

Swan read an article by John Starr, an American living in England, who had patented a lamp with a carbon stick enclosed in a glass bulb with a partial vacuum. Swan set out to improve this by substituting the power-hungry stick of charcoal. A thin filament would require far less electricity to make it glow. He found that carbonised twists of paper, spirals of carbon, worked well.

In 1867 Swan suffered a double blow that put a stop to his experiments: both his wife and his business partner died, leaving him with three children to rear and a pharmacy to run single-handed. It was several years before he returned to the problem of perfecting the light filament. His new approach involved extruding nitro-cellulose to make a uniform and less fragile filament. He had produced an artificial fibre and his methodology led to rayon, the first synthetic fabric. By this time a German working in London, Hermann Sprengel, had invented the vacuum pump, so Swan's filaments lasted much longer. Swan devised a method of hardening them and applying a coating of carbon and a procedure to prevent carbon particles blackening the inside of the bulb. His bulbs could now burn continuously for forty hours.

In 1878 Swan gave several public demonstrations of his light bulb and the following year Mosley Street in Newcastle became the first in the world to be lit by incandescent lights. In 1880 Sir William Armstrong's enormous country retreat in the wilds of Northumberland became the first private home to be entirely lit by incandescent bulbs. The turbine in the nearby stream powered all the lights, the central heating and even the dinner gong.

In 1881 Swan founded the Swan Electric Light Bulb Company. As usual there were detractors. The chief engineer of the Post Office considered electric lighting to be a 'completely idiotic idea'. Instant light took some getting used to. Several hotels reminded guests they 'need not light the bulb with a match'.

Swan was contracted to light the House of Commons in Westminster. The modest scientific dabbler had turned his hobby into a career, but he was about to be challenged by a very different sort of inventor in the United States, Thomas Edison. As a teenager Edison peddled news bulletins and candy on a train. The railway company allowed him to experiment with inventions in the baggage car during the train's six-hour stop at the Detroit terminus. This arrangement was terminated when one of his experiments set fire to the coach. It was not his first conflagration. A few years before, he had burnt down his father's barn. His requirements for inventing were few: just imagination and a warm fire. The fire of ambition made him a formidable rival.

He now had a large barn in the village of Menlo Park in New Jersey; the first major well-equipped research laboratory. His stated aim was to turn out small devices every ten days and something big every six months. He would average sixty-seven patents a year.

Edison's interest in incandescent lights was sparked by Henry Woodward and Matthew Evans in Canada. They had invented a light bulb but couldn't raise the finance to put it into production. Edison promptly bought them out. He set twenty researchers to develop the incandescent lamp and formed the Edison Electric Light Company, although he hadn't yet got a single working bulb. In the search for the ideal filament his team tried hundreds of materials, including fifty types of bamboo and the wiry whiskers of one of the workers. They were swamped with data. 'Results!' he said. 'Why, man, I've gotten a lot of results. I know several thousand things that don't work.'

Edison was a slave driver and he installed an organ to jazz up the staff if they were wilting. He also drove himself, often working day and night and then sleeping for thirty-six hours. He was a great believer in the ten-minute nap, although he even worked as he slept. To solve problems he daydreamed while clutching ball bearings. When he nodded off, his hands relaxed and dropped the ball bearings to clatter on to metal plates placed on the floor. Rudely awakened, he immediately noted down any ideas that came into his mind while on the edge of sleep.

In 1879 Edison invited three thousand people to visit the factory, parts of which were bathed in light. There were only thirty-four light bulbs and most of the illumination was from discreetly hidden oil lamps. In 1880 he patented his light bulb, including the filament that Swan had invented. Swan wrote a letter to the prestigious scientific journal *Nature* refuting Edison's claim to have invented the incandescent bulb. Edison retaliated by suing Swan for patent infringement. His first display of his light bulbs at Menlo Park took place ten months *after* Swan's demonstrations. Swan had taken out patents in 1860 and 1879 ahead of Edison, but they covered only the methodology of manufacturing the filaments. He had refrained from patenting the entire bulb as several other inventors had used and in some cases patented it, so he thought it improper to include their ideas. In this he was unlike Edison, who believed: 'There are no inventions without a pedigree.' Edison claimed that 'Everybody in commerce and industry steals' and openly admitted that he had purloined other people's work. The court found in favour of Swan.

The only option available to Edison was to merge with Swan, but he bristled at having to share the limelight. He was accustomed to putting *his* name on the patent forms no matter which of his now eighty researchers was the inventor. Although the two inventors never met, in 1883 they formed Edison and Swan United. They were not united for long; in 1885 Edison bought out Swan and dissolved their partnership. In Heysham, England, there is a 132-year-old Ediswan light bulb that still works.

Edison's technicians produced a sixteen-watt bulb with a life of 1,500 hours. Edison then did what he did best: aggressive, well-targeted marketing with maximum publicity. It would be a spectacle when he lit up the Wall Street area of Manhattan to impress the richest potential sponsors. It required a power station, 15 miles (24 km) of cables and 800 light bulbs. Edison was given permission to bury the cables underground, but the electricity leaked out and the ground tingled underfoot. This demonstration was followed by the illumination of the New York Stock Exchange and the La Scala opera house in Milan. It was blazingly obvious that incandescent bulbs were the lighting of the future.

AC/DC

Edison had a problem. Power was lost when electricity was transmitted from power station to consumer. If the voltage was doubled the loss could be reduced by seventy-five per cent. Edison was using direct current (DC), which could only deliver low voltage for short distances, necessitating numerous expensive power stations.

There was an alternative. In 1889 Nikola Tesla invented

an alternating current (AC) induction motor (one in which the electricity was generated by magnetism). Unlike DC current, the voltage of AC could be easily increased for transmission and then reduced by a transformer at the receiving end. Moreover AC could deliver over long distances and use thinner, therefore cheaper, wire.

Tesla, a Serbian born in what is now Croatia, arrived in the United States with four cents in his pocket, having been robbed on the voyage. He worked for Edison and improved the design of the company's generators, but Edison refused to pay him the agreed fee, saying his offer was 'just a joke'. Tesla failed to see the humour and resigned, and Edison lost the expertise of one of the most gifted inventors of his generation.

The contrast between the two men could not have been greater. Edison was dishevelled and his modus operandi was to try everything until something worked. Tesla was dapper and more cerebral. With his photographic mind he visualised the induction motor in his head and thought out all the difficulties before building a faultless machine.

To fund his research Tesla sold all his AC patents to George Westinghouse, a businessman who had invented compressed-air brakes for trains when he was only twenty-two. Tesla designed the giant turbines for Westinghouse's new power station harnessing the energy of Niagara Falls. He also showed that AC current could be transmitted to Buffalo, 22 miles (35 km) away, which DC could not have done.

Edison had invested heavily in DC and stubbornly persevered with it. Instead of switching to AC he sponsored a smear campaign to discredit his rival, Westinghouse. Harold Brown, a former salesman for Edison, would show that, unlike the 'completely harmless' DC electricity, AC was a 'death current', and to prove it he electrocuted animals in public displays. He paid street urchins to round up stray dogs. While he was happily zapping them on stage an animal welfare inspector stopped the show. Brown was miffed and assured the man that he had 'enough dogs to satisfy even the most sceptical'. To press home the point he later dispatched horses and calves.

Tesla retaliated with a spectacular display at a meeting of senior engineers, sending thousands of volts through his body and lighting an unconnected lamp just by holding it. The electricity was at such a high frequency that it caused him no discomfort. A photo taken a few years later depicts Tesla sitting nonchalantly in a chair with artificial lighting bolts streaming around him. No wonder he became known as the 'electric sorcerer'.

Brown tried to top that by executing an ageing circus elephant called Topsy. Edison offered the gear to do the job and just in case the electricity wasn't sufficient, the elephant's last meal was carrots laced with cyanide. Poor Topsy was jolted by 6,600 volts and dropped down dead. A film of this 'demonstration' appalled the public.

But Brown was undismayed and invented the electric chair to electrocute humans. The first felon to benefit from this procedure was William Kemmler, convicted of murder. He was first soaked with water and then hooked up to the electricity. It took two attempts to kill him and Westinghouse commented: 'They would have done better with an axe,' although if memory serves me, the axe man took two chops at Anne Boleyn. The electric chair had its admirers. Haile Selassie I, Emperor of Abyssinia, bought three before remembering that there was no electricity in his country. Undeterred, he converted one into a throne. No one denies that electricity can be dangerous. A few years ago I read in the newspaper of Michael Godwin, who was reprieved after being condemned to the electric chair. He subsequently electrocuted himself while sitting on a metal lavatory seat mending his TV set and biting through the live wire. Every year three Britons are killed using their tongue to test if a nine-volt battery works. Thirty-one were electrocuted in the past twenty-two years while watering their Christmas tree with the fairy lights plugged in.

In 1893 Westinghouse's bid to illuminate Chicago's World Fair undercut Edison's bid by half. His brilliant display of 100,000 light bulbs ensured that AC current was the power of the future. Even Edison conceded and switched to AC under licence from Westinghouse. Despite this setback Edison prospered and with almost 1,100 patents became the most successful inventor of all time. He was awarded several honours, but turned down the Nobel Prize when he learned that Tesla would share it. A newspaper poll voted him the greatest living American. When he died electric lighting was switched off for a minute all over the United States.

Tesla was less fortunate, although it must have galled Edison when the Institute of Electrical Engineers awarded his former employee the Edison Medal. Tesla was undoubtedly brilliant, but he was bedevilled by phobias and obsessions. He was afraid of women wearing pearl earrings, obsessed by the number three and by pigeons in the park. At dinner parties he couldn't resist calculating the cubic content of the food on his plate.

When Westinghouse had financial difficulties during the great depression, Tesla generously waived his royalties from the AC patents. In doing so he impoverished himself. When he died in 1943 the only interested parties were FBI men who confiscated all his papers, perhaps looking for the details of some of his wilder projects: the death ray or the results from shooting huge lightning bolts into the sky from a tall tower to attract the attention of extra-terrestrials.

Tesla filed over a hundred patents, including the invention of the fluorescent tube and the vertical take-off and landing (VTOL) aircraft. Although the entire world is now illuminated with his system he has been forgotten by all but physicists. Sixteen years after his death the official unit of electromagnetic induction became the 'Tesla'.

ON LINE

'I did receive fair speechless messages'
— **Shakespeare**

There are occasions when we wish to communicate beyond shouting distance. Throughout history the military developed signalling with drum beats and beacon towers. A chain of flaming beacons reported the fall of Troy. One thousand eight hundred years ago Hadrian's Wall marked the limit of Roman Britain, and in addition to its sixteen forts there were 158 'milecastles' where soldiers stood with their armour rusting in the rain, ready to light a beacon should the Picts threaten. When the Spanish Armada was sighted off the Channel coast in 1558 a string of beacons carried the news to London within twenty minutes.

Giving the right signals

In 1791 Claude Chappe devised signalling machines with pointers and dials fixed on top of towers to transmit information after the signaller had beaten a drum to attract the attention of the recipient in the next tower. The French inventor soon dumped the dials for wooden beams whose relative positions indicated letters or words. It worked well but not if the enemy raid happened when it was foggy or dark. He called his system the *télégraphe* (writing at a distance).

Chappe built a network of communication towers in France and his system was widely adopted elsewhere. In 1796 George Murray, who later became the bishop of the Isle of Man, invented a successful mechanical signaller and was made the first director of telegraphs for the Royal Navy. His greatest achievement was to establish a chain of signal stations stretching from Portsmouth naval base to the Admiralty in London. Seven years later Admiral Popham devised the simple two-arm semaphore that I learned as a boy scout and never used again.

Inventors are not confined to the laboratory or workshop. Stephen Gray lived in a London refuge for 'distressed gentlemen, old soldiers and merchants decayed by pirates'. Clearly Gray was not entirely decayed when in 1729 he discovered that an electric charge could be sent 290 ft (88 m) along a string. He was the first to show that while some materials conducted electricity others did not, and that static electricity could be transferred from one object to another and was lost when the object touched the ground.

Gray set up a bizarre demonstration of his findings for the Royal Society. A boy was suspended horizontally from the ceiling by cords of silk, which material was, he discovered, a good insulator. The lad's foot rested on the wheel of a device producing static electricity. When he gingerly extended his finger towards a tray of metal flakes they became excited by his electrical charge. In another display Gray linked a chain of youths with iron rods and when electricity was applied to one of the youths, all the others became agitated. Gray, *the* pioneer of electrical communication, was awarded the highest honour of the Royal Society. His findings were a harbinger of great things.

Once the voltaic cell had been invented people pondered how to send signals by wire. A difficulty was that an electrical signal faded before it had gone very far. Joseph Henry in the United States and Edward Davy in England independently invented an electrical relay that boosted the signal along the line. Long-distance telegraphy was now possible, but how could an electrical pulse be coded into letters of the alphabet? Pioneers plumped for a separate wire for each letter. The intended letter would be recognised at the receiving end because a metal ball became electrified and attracted the paper below bearing that letter, or the wires were immersed in cups of weak acid and the one with the signalled letter gave off bubbles of hydrogen.

There had to be a simpler way that required fewer wires and less clutter. In 1823 an Englishman, Francis Ronalds, reduced the system to only two wires and set up a working system in the grounds of his house using 8 miles (13 km) of wire. He offered the system to the Admiralty, who gave it the thumbs down: 'Telegraphs of any kind are wholly unnecessary.' As ever, the Admiralty was looking to the future through the wrong end of a telescope.

The breakthrough came from a mismatched duo. William Cooke was a British Army officer who knew little about electricity but had an idea which he took to Charles Wheatstone, Professor of Experimental Philosophy (physics) at London University. In a weak moment Wheatstone had invented the concertina, but he also knew about electrics. Cooke's idea was to use a galvanometer, which measured small electrical currents, to receive and register messages. In

1837 Wheatstone and Cooke patented their telegraph, which had needles that pointed to the letter transmitted. Its signals arrived almost instantly and the operator needed little skill to decipher the message. The only drawback was that no message was private once the operator had deciphered it. Wheatstone and Cooke's system provided a basis for all subsequent telegraphs.

In 1831 Joseph Henry, a passionate man, told his wife to take off her petticoat and she was relieved when he tore it into strips to insulate wires. He was fixing an electromagnet to a transmission wire and the insulation would prevent shorting. Breaking and restoring the circuit switched the magnet off and on, thus releasing and attracting a metal pointer. Stupidly, he didn't patent his invention.

Henry's 'Morse' code

The famous name in telegraphy is Samuel Morse, a portrait painter. He became interested in telegraphy and built a primitive telegraph transmitter from bits and bobs attached to his frame for stretching canvases. Joseph Henry, a fellow American and a skilled engineer, helped him to produce a working system. Morse patented it in 1840, but with no mention of Henry's vital contribution.

Morse entered into a partnership with Vail Ironworks. The agreement was that they would manufacture the devices and the owner's son, a bright young engineer named Alfred Vail, would help Morse. Vail perfected Morse's rudimentary telegraph key for tapping the pulses, while Morse struggled to devise a code that converted key taps into letters. He started by coding common words. One tap for one word, two taps for another and so on. As messages would require a vocabulary of at least thirty words, one word would require thirty taps, clearly a non-starter. Vail took a different tack. He devised a code based on short and long pulses – dots and dashes. He counted the numbers of each letter in a printer's tray and then allocated the shortest codes to the commonest letters. E was just 'dot', whereas an uncommon letter like Y was 'dash-dot-dash-dash', still just four taps on the key. Morse took the credit for the code and Vail, like Henry, was not acknowledged. The receiver printed out an embossed strip of dots and dashes, but the operators soon became efficient at translating the pulses into letters, which they called out to an assistant who wrote them down.

One such telegrapher was the young Thomas Edison. He was out of a job but fate stepped in. His reward for snatching a stationmaster's son from the oncoming wheels of a train was to be taught to operate the telegraph and he was good at it. His telegraph key had a 'tell-tale' device to ensure he was alert. At regular intervals he had to tap a particular letter. Edison rigged up a device that automatically tapped that letter while he read technical books. It worked well until he was rumbled and fired.

Western Union gave him a trial, but his cocky self-confidence must have irritated them because they got their best operator to send a thousand words at top speed. Edison was unfazed and halfway through he tapped back: 'Hurry!' While at Western Union he invented a system that allowed two messages to be sent simultaneously in opposite directions over the same wire. He was so enamoured with the telegraph that he called his first two children Dot and Dash.

The first public telegraph line in Britain was installed in

1839 to control rail traffic between London and West Drayton, 13 miles (21 km) away. Just over a decade later a third of the network was equipped with telegraph wires and telegrams were widely used for business and private communications. In 1845 a suspected criminal was seen boarding a train. His description was telegraphed to London and he was arrested at Paddington station for travelling without a valid ticket. He was later hanged for murder: bad luck for him but good publicity for the telegraph.

In 1851 Morse persuaded congress to fund a telegraph line between Baltimore and Washington, DC. The 40 miles (64 km) of cable was strung on huge telegraph poles with glass doorknobs for insulators. The overhead wires were damned for causing bad weather and ruining the crops. Some farmers even chopped down the poles, but not even an army of axes could halt the progress of the wires. The first transcontinental line was completed in 1861 and by 1864 the phrase 'I heard it on the grapevine' reflected the inexorable spread of the telegraph's tendrils.

Crossing the oceans

If the telegraph were to connect the world, it would have to dip its tendrils into the sea. As early as 1747 Bishop Watson transmitted an electrical pulse under the River Thames in London. A century later Ezra Cornell, who founded Cornell University, laid a cable across the River Hudson, but it was soon damaged by floating ice. Underwater cables needed to be protected, but it would be over a hundred years before a British railway engineer, Thomas Crampton, designed the first securely armoured cable.

River water was one thing but saltwater is a different kettle of fish. It's an excellent earth and the current rapidly leaches away unless the cable is insulated. The best material available was a rubbery, waterproof latex called gutta-percha. The Siemens Company in Germany had observed the making of macaroni and used the same technique to extrude a seamless tube of gutta-percha.

Charles Wheatstone put down his concertina in 1849 to experiment with underwater transmission. He showed that electrical signals could be telegraphed from a boat at sea to a base on shore 2 miles (3.2 km) away. Two years later the SS *Goliath* laid a cable from a telegraph station (a horse box) in Dover across the English Channel to the French station, a bathing machine on a beach near Calais. Within days a French fisherman hauled up the cable in his trawl. Thinking it was a strange seaweed that accumulated gold, he sawed off a length of cable, thus terminating the historic cross-Channel link.

A second attempt used HMS *Blazer* to carry the cable. It was little more than a hulk that had to be towed across the Channel by two tugs. Within 3.5 miles (5.6 km) of the French coast the reel had only 3 miles (4.8 km) of cable remaining. There was no alternative but to lash the cable end to a buoy and return to England for more cable. The completed link worked well for over thirty-five years. In 1853 Charles Bright, who had brought the telegraph to almost every major city in Britain, laid the first successful cable across the Irish Sea to link England to Ireland.

The great challenge was to lay a cable across the Atlantic. By comparison, the previous underwater exploits were merely paddling in puddles. Most people thought it couldn't be done, perhaps because they believed the telegraph worked by tugging at one end to ring a bell at the other. Even supposed experts held that the huge weight of water in the deep ocean would crush even incompressible water, increasing its viscosity, so that any object that sank would not reach the bottom, only descend to the depth at which it encountered water of the same density. Thus bodies committed to the deep would hang for ever at their prescribed depth in a suspended graveyard, as would the proposed cable.

The bed of the Atlantic Ocean was unknown territory; even the depth of the water was uncertain. Depth was measured by lowering a weighted line over the side of the ship, but when the sea floor was a mile or two (1.6–3.2 km) down it was difficult to tell when the weight hit the bottom. Also, currents deflected the line sideways, so it didn't hang vertically, and therefore the length of line paid out was greater than the actual depth. Surveying was laborious, a single sounding in deep water took six hours and soundings were taken at 50-mile (80 km) intervals. Part of the greatest mountain chain on earth runs down the mid-line of the Atlantic and its peaks rise 5,000 ft (1,520 m) above the average depth of water. With cable-laying in the offing a distinguished American oceanographer named Lieutenant M.F. Maury collated the existing depth data and instigated more systematic surveys. He suggested that the direct route from Ireland to Newfoundland traversed the mid-Atlantic ridge at a less mountainous area, later named Telegraph Plateau.

At that time the speediest communication between Europe and the New World was a ten-day voyage. The cost of a telegraphic link would be substantial but the rewards of rapid communication would be colossal. The initial driving force was American businessman Cyrus Field. He came to England to excite financiers with the romance of an underwater cable and he succeeded. A meeting in Liverpool attracted numerous subscribers including novelist William Makepeace Thackeray and Lord Byron's widow. So Field immediately ordered 2,500 miles (4,025 km) of armoured cable. The fine wires from which the cable was made would have encircled the globe thirteen times. It was manufactured in lengths of 2 miles (3.2 km) and these were spliced together.

Field also persuaded the governments of Britain and the United States to provide vessels to deploy the cable. The USS *Niagara* was the newest ship in the US Navy and her captain wept as she was radically modified for cable laying. Not to be outdone, the British Admiralty released HMS *Agamemnon*, a fully rigged, geriatric steamship that creaked at the very thought of toting 1,436 tons (1,460 tonnes) of cable across the Atlantic.

In 1857 a small flotilla of ships left Valencia, an island off south-west Ireland, after attaching the end of the cable to the land line of the telegraph station (a small hut). The cable was so thin that they had travelled only 4 miles (6.4 km) out to sea when it snapped and had to be repaired. A survey vessel went on ahead, making depth soundings, and the *Niagara* steamed behind at the pace of a sleepwalker, with the cable easing off the reel. On the third day the depth tripled and the strain on the cable increased to 1 ton (1,016 kg). The cable broke and sank 12,000 ft (3,660 m) down to the sea floor. The crew were heartbroken and the ships turned around and headed for England.

The next year they tried again, this time with a heavier-gauge cable and a machine for paying out the cable that responded to the strain it was under. They also had a much more sensitive galvanometer, invented by William Thompson (later Lord Kelvin), with which they could continuously test whether the cable was still conducting electricity. The plan was for two ships, each carrying half the cable, to steam to the middle of the Atlantic and there splice their cables together. They would then depart in opposite directions paying out the cable; the *Niagara* heading for Newfoundland and *Agamemnon* returning to Ireland. That way the task would be completed in half the time.

A storm blew up, which was bad news for Samuel Morse, who was aboard the *Niagara* and had been seasick before they left port. It was worse for those on *Agamemnon*, whose deck planks were gaping by up to 1 in (25 mm), allowing water to leak into the cabins. A wave crashed through the windows of the wardroom, throwing officers across the floor.

Both ships survived the storm and the link was consummated. It was acclaimed as a triumph on both sides of the Atlantic. Cartoonists depicted King Neptune hanging out his washing on the cable. New York was ablaze with lights and City Hall was especially brilliant as its dome caught fire. The surplus cable was made into souvenirs and US President James Buchanan and Queen Victoria exchanged congratulations by telegraph. Her Majesty's message of only ninety-eight words took over an hour to transmit because the signal was so blurred that it was difficult to distinguish the dots from the dashes and each phrase had to be repeated several times. The cable wasn't fully watertight and on the evening of a banquet in New York to celebrate this great achievement the line began to fade and die. But the dream lived on.

Eight years later there was a new attempt using a better

armoured cable whose seven strands of inner wiring were of the purest copper to enhance transmission. They were three times thicker than those in the previous cable. This time the leading force was Daniel Gooch, an English engineer who had designed 340 railway locomotives, the fastest and safest of the age. He became the director of the Great Western Railway and was on the board of the Telegraph Construction and Maintenance Company.

The first decision was to choose the ship. Brunel's *Great Eastern* was an engineering triumph but a commercial failure. It was auctioned and Gooch bought her for a knockdown price. Brunel's 'great babe' was a leviathan, five times larger than any other vessel. It was the only ship capable of carrying *all* the transatlantic cable. The 2,300 miles (3,700 km) of cable was coiled into circular tanks each 50 ft (15.2 m) in diameter and 20 ft (6 m) deep. They contained sea water so that the submerged cable could be tested as if it were under the sea.

In 1865 the *Great Eastern* steamed to Valencia to connect the cable to that from the land station. Sir Robert Peel, the son of the famous statesman, gave a speech: 'We are about to lay down at the very bottom of the mighty Atlantic . . . this silver-toned zone to join the United Kingdom to America.' Day after day passed to the rhythm of the machine paying out the cable. The ocean was a well-behaved world of gentle swells, with the occasional whale taking a breath and casting a glance at something even bigger than itself.

It was too good to last. Gooch described the setback in his diary: 'What a change a few hours may make . . . last night I went to bed perfectly satisfied all was going as well as it was possible for anything to do, but at three this morning

I learned the cable had become defective.' A length of cable had to be reeled back on board and on close examination they found a piece of wire from the armoured sheath had been forced all the way through the cable. It could have happened accidentally when the cable was passing through the paying-out machine, but when the same thing happened again it raised the possibility of sabotage. Thereafter, only the most trusted personnel were allowed into the cable storage tanks and the uncoiling was closely supervised. Gooch was on the rota of supervisors: 'It was anything but an agreeable job to stand for two hours in a tank of cable covered with tar and like a girl with a skipping rope, jump over the cable as it runs out, say every minute.'

Sabotage or not, there were no further incidents until: 'All is over . . . our cable is gone . . . The cable broke a few yards from the ship . . . This is indeed a sad and bitter disappointment. A couple more days and we would have been safe.' The cable lay 2.5 miles (4 km) below the surface. The crew deployed a five-pronged grapnel and dragged it over the bottom of the sea. They hooked the cable three times and tried to haul it up. On one attempt the cable rose from the waves like a sick sea serpent and when almost within reach it slipped back into the deep. There was nothing to do but mark the site with a buoy and return home.

The following year the *Great Eastern* set sail again, laying out a new cable. The only problem was that the heavy cable kept getting tangled and was the very devil to sort out. Within fourteen days they arrived at their destination, the aptly named Heart's Content Bay in Newfoundland. The escort ships returned to the buoy that marked the cable lost the previous year. It took three weeks to recover it, splice it to a new cable and join it to Newfoundland. There was not just one cable linking the two continents: there were two.

It was a magnificent achievement for Gooch and his colleagues, and the redemption of the *Great Eastern*, but her time in the spotlight was brief. She ended up beached in the River Mersey estuary as a giant billboard for Lewis's department store in Liverpool. She didn't die easily; it took two hundred men two years to demolish her. Even today a few of her iron plates remain on the shore of the Mersey.

Valencia became a centre for world communications. With the new cables it took just over a second for a signal to traverse the Atlantic. But tempers rose in the Newfoundland station if there was not an instant response. Valencia's superintendent explained that only George and Mackey were on duty and 'The former was attending a fire in the kitchen, the latter a call of nature in a field.' Soon afterwards the Valencia station moved into grand new quarters with a lavatory. The telegraph operators prided themselves on being fast and accurate. On one occasion the receiver signalled 'Stop'.

The response was, 'Why?

'Can't read it.'

'Cant read what?' was the tetchy reaction of the operator, thinking his skills were being criticised.

'My own handwriting,' came the reply.

In England the telegraph system was acquired by the Post Office and a telegram could be sent for sixpence. Mr Scudamore was put in charge and his remit was to establish telegraph stations all over the British Empire. By the 1880s almost every country was connected to the world by telegraph. Scudamore managed to spend vastly more than the several million pounds allocated to him. When challenged by

Parliament he replied: 'How do you expect a canary to lay an ostrich egg?'

Even today every piece of information sent between London and New York – financial transactions, emails, Facebook updates and bids on eBay – travels through an underwater cable.

WIRED FOR SOUND

'One day there will be a telephone in every
major city in the USA'
– Alexander Graham Bell

The success of the electric telegraph was short-lived, for receiving printed messages did not match conversing. Several people were convinced that if electric pulses could be sent down a wire, so too could sounds. In 1854 French scientist Charles Bourseul wrote an article on the transmission of speech which inspired Johann Reis, a German professor of physics, to build his Telephon. In this device an electrical current set a membrane vibrating and sending a signal to vibrate a membrane in the receiver. This was the principle behind future telephones. Reis's machine worked fine for sounds but duplicating the complex wave form of human speech was beyond him.

In 1835 an Italian theatre electrician, Antonio Meucci, emigrated to Cuba. To supplement his salary he practised electrical therapy, a fashionable cure-all. One day, having attached the electrodes to a patient and retired to the next room to turn on the electricity, Meucci heard the client cry and to his astonishment the voice came through the wire.

When the theatre folded he moved to New York and took

up inventing. He spent fifteen years perfecting his Telletrofono. By all accounts he had a working telephone link between his home and workshop. In 1871 he filed a caveat at the patent office giving notice of an impending patent submission. A caveat is much cheaper than a patent and must be renewed annually.

Meucci was on the brink of becoming rich and famous when tragedy struck. The boiler on the Staten Island ferry exploded, killing 120 people and badly scalding many others, including Meucci. Without an income and with hospital bills to meet, his wife sold most of the models of his inventions and he couldn't afford to renew his caveat.

His sponsors deserted him, so he approached Western Union and gave them documentation and a prototype of his telephone, hoping they might take up his invention. They didn't respond and when he pestered them he was told that his prototype was lost, and with it his chance to commercialise the telephone.

Telephone Bell

Instead, the fame went to a Scot, Alexander Graham Bell, who came from a long line of speech therapists. His grandfather had trained a dog to talk by getting it to growl and then shaping its mouth by hand. Alexander's father, Melville, a speech therapist, was the inspiration for George Bernard Shaw's Professor Henry Higgins who taught Eliza Doolittle to talk posh.

As a child and an aspiring pianist Alexander discovered that if he pressed a piano key, the strings of a piano in the next room reverberated faintly *in the same note*. This may have reverberated in his mind years later when he became interested in telephony.

His family emigrated to Canada and Bell devoted himself to speech tuition and was given a post at Boston University teaching the deaf to talk. He would later teach Helen Keller, who had been deaf and blind from infancy. With his help she became a writer.

Bell also began experimenting with a means to send multiple messages down the same telegraph line simultaneously. His 'harmonic telegraph' had matching sets of vibrating tuning forks sending messages, each on a different musical tone. Bell was no electrical engineer, so he hired a bright young man named Thomas Watson. In 1875 they patented the harmonic telegraph and then turned their attention to the telephone.

At the Massachusetts Institute of Technology Bell had seen the Phonautograph. It was a device on which the sound wave of the human voice was made visible by carbon powder on a metal sheet. He realised that to transmit speech the sound waves would have to be modulated to match the human voice. Johann Reis and others had used an on/off current producing a digital signal when what was needed was a continuous modulated current, an analog signal.

On St Valentine's Day in 1876 Bell filed a patent for the telephone even though he didn't have a working model. This was rectified within days when, according to legend, he spilt battery acid down his trousers and called for help: 'Mr Watson, come here. I want you.' Presumably the expletives have been deleted. Watson dashed upstairs shouting: 'I could hear your voice!' The prototype worked, so as soon as Bell could change his trousers a second patent was filed.

Bell had previously refrained from lodging a caveat as he had promised his sponsor that he would not do so until the sponsor had filed a patent in London. This gentlemen's agreement almost cost him his priority, for only hours after Bell filed his first patent in the USA, an Ohio inventor named Elisha Gray filed a caveat for a similar and superior telephone.

By this time Bell was so strapped for cash that he offered George Brown, a Canadian publisher, exclusive British Empire rights to his telephone for only $150. Brown said it wasn't worth it and thus declined a huge share in the most lucrative patent ever granted. Bell also offered to sell his patent to Western Union Telegraph, but their evaluation was: 'The telephone has too many shortcomings to be seriously considered as a means of communication. The device is inherently of no value to us.' They were in a great position to commercialise the telephone as they had a telegraph network that could carry phone calls. Indeed they soon founded a subsidiary telephone company and hired Elisha Gray so that they could use his rights and his telephone. Bell sued for patent infringement and had the advantage of a prior patent and a good lawyer. Western Union's advocate claimed that Bell had been shown Gray's papers by a patent clerk, which accounted for the improvements in Bell's second patent. The judge believed Bell's denial on oath, so Western Union settled out of court and accepted twenty per cent of Bell's telephone rental revenues. Gray lodged numerous law suits, none of which was upheld.

An advertisement for Bell telephones assured customers that they required no skilled labour or technical education. Perhaps if he had been more technically educated his machines would not have had such a faint signal that callers had to shout – as many mobile-phone users do today. David Hughes in London invented a device that could pick up the sound of a fly's footsteps. His 'microphone' was incorporated into the telephone and has since assisted singers who had weak voices.

Thomas Edison fine-tuned the transmitter and made the telephone more user-friendly. Bell's original mouthpiece was also the earpiece, so the caller had to continually shift it from mouth to ear. Indeed the instructions read: 'Do Not Listen with your Mouth and Talk with your Ear.' Better-off customers bought two phones to solve the problem.

A telephone is no use unless you can connect with the person you wish to call. The first telephone exchange was installed in New Haven, Connecticut, in 1878. The caller's line had to be manually plugged into the socket for the recipient. At first the caller had to tap on the telephone's diaphragm in the hope of attracting the operator's attention. The operators were boys until they proved to be unreliable and cheeky and were replaced by young women. The girls had to be single, alert and at least 5 ft 3 in (1.6 m) tall so that they could reach the plugs at the top of the switchboard. One telephone company advised its customers to hail the operator with a hearty 'Ahoy! Ahoy!' There was some debate about how the operator should greet the caller. Their usual 'Who are you?' or 'Are you there?' were not welcoming, so Edison invented a new word – 'hello' spelt with an 'e' – and it caught on. The operator usually knew when the call had finished, because she was listening in.

Almon Strowger, a Kansas undertaker, didn't trust the operators and was convinced they were diverting his calls to enrich a rival mortician. His paranoia drove him to invent a 'girl-less, cuss-less, out-of-order-less, wait-less telephone'. Although he was no mechanic he built a model of his idea with pins, a pencil and a circular box for detachable collars. Believe it or not, it represented an automatic telephone switch-board. In the finished device pressing buttons in the right sequence sent pulses that told a drum how far to turn to reach the correct wire. Once the telephone companies switched to automatic exchanges operators became redundant except for long-distance calls.

While Australian farmers with no transmission lines hooked up their telephones to their wire fences, which gave reason-able reception over a short distance, Edison planned on a grand scale. In 1885 he predicted: 'It is probable that by means of repeater stations communication can be had over all parts of the United States.' How right he was. Soon every city was cobwebbed with overhead wires. In 1915 the first transcontinental call was made by Bell on the east coast to Watson on the western seaboard. Bell's words were once again: 'Mr Watson, come here. I want you.'

Laying telephone cables across the Atlantic was a greater challenge and came much later. As usual there were doubters who were sure the pressure would squeeze the cable and the caller's voice would be squashed to a mouse's squeaking. In reality the problem was that the signal faded over long distances. So the first transatlantic telephone cable was not laid until 1956. It needed fifty boosters to get the signal across. Even then the cable could carry only one conversation at a time and a three-minute call cost double the average weekly wage.

Bell's American Telephone and Telegraph Company became the biggest and most profitable corporation in the United

States, making him and Watson very rich. Watson filed sixty patents of his own, including the telephone bell that rang when somebody called. He retired young and became a successful actor in England, where his Shakespearean roles garnered good reviews.

Bell continued to invent and devised a metal detector to locate a bullet in President Garfield after an assassination attempt. However, the metal bedsprings confused the detector and he failed to save the president. Among his successful inventions was an iron lung to artificially ventilate patients who couldn't breathe for themselves and, with collaborators, a hydrofoil that broke the world water speed record. His Photophone transmitted sound on a beam of light and was the precursor of fibre optics used today in telecommunications. With the profits from his inventions he established what is now called the Alexander Graham Bell Association for the deaf. He also co-founded *Science,* one of the finest scientific journals in the world, and the National Geographic Society. The decibel, used to measure the level of sound, was named after him.

In his later life Bell was in awe of the telephone: 'I often wonder if I really invented it, or was it someone else I had read about?' He admitted that he never really understood the principles of electricity and couldn't grasp that someone in Washington could converse with someone in Paris. On his death all the phones in the United States failed to ring for a minute.

The phone took much longer to gain acceptance in Europe than in the USA. By 1910 the United States had seven million telephones. Europe would take fifty years to reach that figure. In Britain the authorities that should have embraced the new technology dismissed it. The head of the General Post Office (GPO) insisted that telephones were unnecessary as 'there are plenty of small boys to run messages'. The GPO's chief engineer admitted: 'I have one [telephone] in my office, but more for show. I do not use it because I do not want it.'

The telephone also had a dark side. It was well known that public phones were 'an agent of contagion'. This belief led to a flurry of patents for means of disinfecting them. The Hygiephone guaranteed to combat the dangers resulting from the 'promiscuous use of telephones by diseased people', who have 'breathed, smoked or coughed into, or held against the perspiring ear of a person of doubtful cleanliness'. In recent times we have become complacent regarding the peril of the sweaty ear. One invention involved a gas flame that ignited as the earpiece was replaced on its hook. It cremated germs and melted the ear of the next user.

Modern telephones

I remember when telephones were chunky black devices so heavy that if one fell on your foot you couldn't foxtrot for a month. The earliest portable phones could easily have been mistaken for a shoe box until you felt their weight – the original Vodaphone was affectionately known as 'the brick'. The first palm-sized radio phone was developed by the Swedish company Ericsson in 1979. It could send and receive calls from a local 'cell'. The call would then be relayed from cell to cell in a network.

An earlier network serviced Rabbit cordless phones. The phone and calls were cheap but it was not entirely satisfactory because it could only *make* calls, not receive them.

It was also an analog system and was soon superseded by digital, two-way cell phones. The phone became truly mobile. In 1992 Sir Edmund Hillary in New Zealand received a phone call from his son who was standing on the summit of Mount Everest.

In 2007 the mobile phone came second behind 'weapons' in a poll of the worst inventions of all time. Nonetheless, surveys in 2011 revealed that texting was the most popular means of communication among adults. They averaged 200 texts a month. On a typical night out they spent forty-eight minutes on the phone, sent three emails, twelve texts and several photographs (probably of their tipsy companions). Most of those questioned admitted that they would be desolate without their mobile phone and *had* to have it within arm's reach sixteen hours a day. One in five took it to bed with them. How often they misused the vibrating mode was not mentioned. Doctors now have to treat new forms of repetitive strain injuries such as 'iPod finger', 'Blackberry thumb' and 'text neck'. It reminds me of Thoreau's claim: 'We have become tools of our tools.' Today many inventors slave away to create devices designed to enslave their user. Beware of geeks bearing gifts.

ON THE CREST OF A WAVE

'Radio [communication] has no future'
– Lord Kelvin, President of the Royal Society, 1899

The public marvelled at the ability to send messages long distances through cables. But when Heinrich Hertz at the University of Kiel predicted that it would be possible to 'transmit intelligence without wires' most scientists thought it was telepathy, a piece of electrical quackery. Hertz's laboratory was alive with the crackle of electrical sparks and he had bent a zinc sheet into the shape of a parabolic mirror. It was the ancestor of the modern dish antenna used for transmitting radio waves and that was exactly what he was doing. The electric spark propagated invisible waves in the air, like the ripples on a pond created by a stone. In 1887 he detected these electromagnetic waves 66 ft (20 m) away from the source.

Hertz was only thirty-six years old when he died and didn't live to see his prophecy fulfilled. In 1894 Professor Oliver Lodge of Liverpool University gave a memorial lecture to Hertz at a meeting of the British Association for the Advancement of Science. He amazed the audience by transmitting Hertzian waves and receiving them with his detector at the far end of the lecture theatre. It was the first proof they existed and it opened a new world of electricity, because at a particular frequency these waves could carry sound. Lodge believed they wouldn't travel very far.

Marconification

The man who proved him wrong was Guglielmo Marconi. His father was a successful Italian businessman and his indulgent Irish mother was a member of the Jameson whiskey family, so Marconi had the means to experiment. The butler was his laboratory assistant and an old telegraph operator taught him to tap out a message in Morse code. He was soon transmitting signals to a receiver he had built. His most significant achievement was to send a signal to his brother who was out of sight behind a hill 4,500 ft (1,370 m) away. Clearly the waves could travel over topographical barriers.

The Italian Ministry of Post and Telegraphs thought Marconi's invention was no improvement over cable telegraphy. The Italian admiralty also turned down the invention that would have given them an advantage over every navy in the world. In 1896 Marconi took his devices to England. He arrived just after an anarchist had exploded a bomb in the London Underground Railway, so the customs officers considered all foreigners suspect. When they saw all the electrical equipment in his luggage their inspection of it was brutish, falling little short of whacking it with a hammer.

His patent application was therefore for the principle of wireless telegraphy, though he assured the patent clerk that he had invented the devices too. He brandished a box which he kept securely locked because all it contained was a battery, a tube of iron filings and a bell. The patent for radio telegraphy transmission was granted to Marconi, who was only twenty-three years old. He fooled the patent office, and the judgement of an English inventor and xenophobe was: 'No Italian or other foreigner was ever really fair in their judgements so it is quite unreasonable to expect them to be so.'

Fortunately, William Preece, the Consulting Engineer for the General Post Office, was impressed and provided technical help to improve the devices. Preece tried to persuade the GPO to buy Marconi's patent. He estimated that an offer of £10,000 (around £1.1 million today) would be irresistible. However, the Jameson family had mustered a group of financiers who offered a far bigger lump sum plus sixty per cent of the company's shares. Preece was angry that Marconi had forgotten all the help he had rendered to perfect his invention. Preece, who was said to be 'dead against Marconi', rejected his request for a licence to handle messages from GPO offices. Having been denied the lucrative telegram business, the newly established firm of Marconi Wireless Telegraph concentrated on selling its equipment to shipping companies as ships had no means of long-distance communication. The Admiralty chose Marconi to fit wireless sets in ships and shore stations, but other contracts were slow in coming. To publicise his system Marconi transmitted information from the yachts taking part in the America's Cup race, and successfully sent a Morse-code message across the English Channel.

Not all the publicity was favourable. An influential journal accused Marconi of theft as he had not credited Oliver Lodge's prior work. To restore confidence a lecture was organised at the Royal Institution in London which would culminate in a wireless transmission from the Marconi station in Cornwall. The event did not go to plan. Just before the transmission was due, a Morse message came through which translated to:

'There was a young fellow from Italy
Who diddled the public quite prettily.'

It was a disaster because Marconi had boasted that his transmissions were absolutely secure, but he had fallen foul of the man who invented hacking. The culprit was a well-known magician called Nevil Maskelyne who had been hired by the cable companies to eavesdrop on Marconi. He had erected a radio mast not far from the Royal Institution and tuned into Marconi's frequency. Marconi's claim was shattered.

Marconi was desperate to restore public confidence, so he made a bold decision to try to send a message over the Atlantic. The greatest distance he had transmitted was only 155 miles (250 km), it would take a far more powerful transmitter to throw a signal 2,400 miles (3,860 km) from Cornwall to Newfoundland. Graham Bell and Lodge dismissed it as impossible. Their reasoning was that radio waves travelled in straight lines and could not bend to follow the curvature of the earth. They couldn't know that the waves could reflect down from the ionosphere.

To achieve the greatest range a transmitter must be as high as possible. So Marconi built his transmitting stations on wind-whipped promontories above precipitous cliffs, using masts that were 200 ft (61 m) tall. Not surprisingly, the masts in Cornwall and Newfoundland were toppled by storms and had to be rebuilt.

Clearly a puny spark would not propel a radio wave across the wide ocean. Dozens of generators were used to produce a middling-sized lightning bolt. It banged like 'explosions of a Maxim gun' and the station was known as 'the thunder factory'. The Morse key would probably have melted under this assault, so a heavyweight lever was used instead.

On the big day a gale buffeted the Newfoundland station. To get the maximum height a large balloon was launched on 600 ft (180 m) of wire acting as the antenna, but the balloon broke away and vanished into the dark sky. The next day there was a fierce winter storm with horizontal snow. The operators deployed a kite that rose to 400 ft (120 m), where it juddered and swung wildly, but the wire held.

Marconi was below in the station listening through earphones for hours, awaiting the agreed signal. It came faintly as if exhausted by its long journey: 'dot-dot-dot' and then again: 'dot-dot-dot'. It came through seventeen times and Marconi was elated: 'I knew that I was absolutely right . . . The electric waves . . . had traversed the Atlantic serenely ignoring the curvature of the earth.'

But he had failed to have an independent official confirm what had happened and even more surprising he had not switched on the printer which would have delivered written testimony. This left scope for his critics, and Edison said: 'I don't believe a word of it.' Lodge and Preece suggested that the signals could have come from atmospheric interference, a nearby ship or a prankster.

Nevertheless, the cable wireless companies were concerned. One threatened to sue Marconi for breaking their legal monopoly of transatlantic telegraphy and demanded that he close down his transmission stations. It was an idle threat, for Marconi's main worry was rather that the transatlantic link was by no means reliable. It worked only at night and as dawn broke it failed and sometimes it didn't work at all for days on end. Marconi adjusted the antenna – until the gales adjusted it catastrophically. He built a new station near Clifden, on the west coast of Ireland, as it was closer to Canada than the Cornwall transmitter. The new station generated a colossal 300,000 watts, but to no avail. The secret was to use a short-wave frequency as it travelled much farther on less power. Marconi's new station was several hundred times more powerful than necessary and exorbitantly expensive. He was rapidly sliding towards bankruptcy. Then he had a great stroke of luck – a homeopathic doctor had murdered his wife in London.

Hawley Harvey Crippen had scarpered with his mistress, Ethel Le Neve, after his wife had mysteriously departed. On the day after she vanished Crippen was reportedly his 'usual cheerful self'. He was said by all to be kind-hearted and gentle and he and his wife never rowed. The runaway couple boarded the *Montrose*, bound for Quebec, but there was no Crippen or Le Neve on the passenger list.

Meanwhile back at the Crippen residence the police found something nasty under the cellar floor. It wasn't exactly a body as the flesh had been removed from the bones and the skin stripped away and laid beside the remains in folds like a fallen curtain. On the *Montrose* Captain Kendall radioed to the Marconi station in Cornwall his suspicion that two passengers who had embarked as Mr Robinson and his sixteen-year-old son were the fugitive Crippen and his mistress. The next day Chief Inspector Dew boarded the *Laurentic*, a faster ship than the *Montrose*.

It was an exciting pursuit, with Captain Kendall transmitting daily dispatches to the newspapers. Millions of people were following the chase, and only Crippen and Le Neve were unaware of what was happening. The fact that a murderer had been snared by an invisible net of airwaves and the public had been privy to every twist and turn established the credibility and magic of wireless telegraphy.

Marconi's business began to boom and he became a celebrity. He was difficult man, patient yet bad-tempered, shy but longing for the limelight. He fell head over heels for Beatrice O'Brien, daughter of Baron Inchquin. On their wedding night he surprised his bride by taking a dozen clocks from his valise. Each displayed the local time in various cities. She was left in no doubt that his work came first. Indeed she spent her 'honeymoon' on a bleak, snowbound cliff in Canada while Marconi sorted out the transmission problems at his station. The marriage was doomed, and even when he lost the sight in one eye in a car crash the remaining one could still wink and women were drawn to him.

Not actually a great inventor, Marconi resorted to trial and error and depended on the inventions of others. When he was granted the first patent for wireless telegraphy he had nothing beyond what Lodge had demonstrated two years before. Marconi's lawyer advised him to patent everything, which he did, whether he had first invented it or not. In 1901 Lodge challenged Marconi's latest patent for a means of tuning wireless transmissions as it was similar to Lodge's patent filed four years earlier. The court agreed. Marconi's achievement was to focus on the commercialisation of wireless telegraphy and doggedly persevere despite setbacks. Wireless telegraphy was just the start of an electrical revolution. Electromagnetic waves had many more uses up their sleeve.

Searching the skies

Wars are good for inventors. Governments suddenly feel a need for scientists and new devices of destruction or defence. During what was called the Great War, as if all previous conflicts left something to be desired, aeroplanes were used for the first time. Britain suddenly realised that even an island was vulnerable to air attack. Early warning of advancing enemy aircraft was essential, so spotter stations were set up on the coast. Boffins believed that planes might be heard before they came within sight. To collect and concentrate sound a Major Tucker designed and constructed gigantic 'sound trumpets' with curved concrete walls. With these, planes could be detected up to 20 miles (32 km) away, which gave the British fighter pilots about ten minutes' warning. But as aircraft became faster enlarged ear trumpets were hopelessly inadequate.

In between the wars several countries experimented with new methods of detection. Long ago Hertz had shown that radio waves were reflected from a metal mirror and in 1900 Tesla proposed tracking moving ships at sea by reflected radio signals. In 1922 US Navy engineers reported that when a ship passed between a radio telegraph transmitter and the receiving station the radio waves were reflected. This spurred the Naval Research Laboratory to begin experimenting with electromagnetic waves. Nine years later, at the General Post Office in England, engineers noticed distorted radio telegraph reception when an aeroplane flew by.

The darkening clouds over Europe stimulated the British government to consider the possibility of war. The Air Defence Research Committee, perhaps after watching too much *Flash Gordon*, asked scientist Robert Watson-Watt if a 'Death ray' that would paralyse planes in flight was feasible. He told them it was not, but the canny Scot softened their disappointment by outlining a scheme to detect incoming aircraft at a distance. His idea was to send out radio waves which would be reflected, just like an echo when shouting into a cave. The silent 'echo' could be displayed as a blip on a cathode-ray tube, much as a heartbeat is seen on a medical monitor. The time it takes to return gives the distance and a movable antenna reveals the direction. As the signal returns within a tiny fraction of a second, new equipment had to be devised to process the data.

Using an existing transmitter tower, Watson-Watt attempted to spot a flying aircraft 8 miles (13 km) away. In his own words: 'It was demonstrated beyond doubt that electro-magnetic energy is reflected from the metal components of an aircraft's structure and that it can be detected.' The Committee was convinced and a custom-built transmitter showed it could detect all aircraft within a 40-mile (64 km) radius. Watson-Watt's team soon realised that pulses of short wavelengths (microwaves) gave better resolution and greater range. This new system was named RADAR, an acronym for Radio Detection and Ranging.

Radar was given the highest priority and by 1938 a chain of early-warning stations, able to detect aircraft 100 miles (161 km) away, stretched from London to the English Channel and along the entire Channel coast. Watson-Watt had not only made radar practical, he had invented a system that determined altitude, speed and direction of incoming aircraft even in the thickest fog and on the darkest night.

Watson-Watt's patents were kept secret until 1947. The Luftwaffe used a system of radio beams to guide their bombers to targeted cities. British scientists found a way to deflect the beam so that many bombs devastated remote fields instead of towns. The German aircrews suspected that something was amiss but were reluctant to inform their commander-in-chief, Hermann Goering, as he believed the beams were infallible. The Battle of Britain was a close-run thing for the outnumbered fighters of the Royal Air Force, but thanks to radar the ability to deploy aircraft to maximum effect swung the balance in their favour. To conceal why RAF pilots were so successful at combating bombers in night raids the Air Ministry circulated a rumour that British pilots ate bushels of carrots to improve their night vision.

One of the most fruitful collaborations of the war was that between scientists working on radar. The speed at which new ideas were turned into operational devices was breathtaking. The Americans contributed their expertise in microwaves, as well as the now familiar circular screen swept by a beam like the racing finger of a clock. The British provided the magnetron, a highly efficient device that generated the pulsed microwave radiation essential for radar. A team at Birmingham University invented a transmitter that was as powerful as those used in radar stations, but was small enough to slip into your pocket. This enabled radar to be installed in the cockpit of aircraft to scan the ground for navigation and target-seeking.

The range of radar signals in modern warplanes is over 2,000 miles (3,200 km) and radar is used to guide weapons to their target. The latest reconnaissance drones process pictures of objects from several angles to produce a high-resolution 3D image. Software can even erase forest canopies to reveal what lies beneath.

Sounding the seas

In 1915 the Admiralty set up a board of Invention and Research with a view to harnessing the skills of inventors and scientists for the war effort. Perhaps inadvisably, the public were invited to send in their suggestions The success of U-boats in the Atlantic had rattled everyone's confidence, so several people wrote in with ideas for detecting submarines underwater. One correspondent suggested holding a divining rod over a sea chart; another recommended a squadron of eager seagulls skilled at spotting periscopes.

The Royal Navy was reliant on hydrophones, waterproof microphones that dangled over the side of the boat. A sailor with earphones listened for the sound of a submarine's propeller. The drawback of this was that the boat had to be stationary otherwise its engine would drown any other sound, and an immobile vessel was a sitting duck. The sound the sailor heard was likely be the whirr of an inrushing torpedo.

Hence a selection of Navy brass and scientists assembled at a public swimming bath in London to meet Joseph Woodward and his divers – Barker and Jumbo. Woodward had boasted that he could train his performing sea lions to lead anti-submarine boats to their prey. The session in the swimming pool convinced the audience that the sea lions had more sensitive ears than a hydrophone. Woodward claimed that within two weeks he had taught Queenie, a placid resident of London Zoo rather than a seasoned performer, to come to the sound of a buzzer. Trials in a Welsh lake also went well. Just as the Admiralty considered commissioning Barker and Jumbo into the Royal Navy, they failed miserably in sea trials. Faced with the delights and distractions of the ocean, they forgot all about buzzers and submarines. Surprisingly, in the 1960s the US Navy trained dolphins to lay mines on submarines and recently sea lions have been conscripted to guard warships against saboteurs.

In 1915, however, they didn't cut the mustard. Fortunately, French scientist Paul Langevin propagated ultrasonic waves underwater and detected the echo from an iron sheet 328 ft (100 m) away. Several million years earlier bats had harnessed ultrasound signals to avoid colliding with objects in the dark. With the help of British and American experts Langevin produced an active underwater detection system that the US Navy dubbed SONAR (Sound, Navigation and Ranging). Unfortunately, the system was not perfected until the armistice, but it was vital during the Battle of the Atlantic in the Second World War. Submarines became so quick at taking avoiding action that hydrophones are now towed behind ships or are permanently anchored on the sea floor, automatically reporting back submarine movements.

The most unexpected use of radar was for 'viewing' the heavens. In 1931 Karl Jansky, an engineer at Bell Telephone Laboratory, was told to trace the source of interference in radio transmissions. To his astonishment it was coming from the constellation of Sagittarius in the Milky Way. The company shrugged its shoulders and tried to find ways to filter it out. It was left to Grote Reber, an amateur astronomer in Illinois, to build a dish and antenna to search the heavens for radio waves. He found them almost everywhere and he mapped the new universe of sound. In 1946 two Australian engineers, John Bolton and Bruce Slee, working at a radar station in Sydney started a secret project of their own. They 'borrowed' bits and pieces of redundant equipment and began digging a depression in a sandy area nearby. They dug in their lunch break, and at night like grave robbers, until three months later they had excavated a bowl-shaped depression. To make this 'dish' reflect on to the central antenna they laid metal ties from packing cases on its upper surface. They had constructed the second-largest radio telescope in the world. It couldn't be pointed at a particular region in space but the rotation of the earth enabled it to scan the sky. They heard for the very first time hidden stars where optical telescopes saw only cosmic dust. Only then did they let their boss in on their secret. He funded them to build an even bigger dish, 80 ft (24 m) across, with which they pinpointed the very centre of our galaxy.

It was not until after the war that professional astronomers began to build huge radio telescopes like the 250-ft (76 m) diameter steerable dish at Jodrell Bank in England. Sir Bernard Lovell, the man behind the great dish, was convinced that Russian agents had tried to poison him on learning that his telescope could detect a missile launch and track its course. The Jodrell Bank dish, now called the Lovell Radio Telescope, became the world's early-warning system.

The recent trend is not to have ever-larger single dishes but numerous smaller ones. Astronomers can then combine the incoming data from all the dishes to create a much higher-resolution picture of the hidden universe. Because signals are absorbed by water vapour the arrays of radio telescopes are situated in arid regions. The latest plan is to erect three thousand dishes in what is called the Square Kilometre Array because the total area of all the dishes adds up to 1 sq km (0.39 sq miles). Australia and South Africa will each house half of the array, which is expected to be completed by 2024.

Despite its military origins radar has insinuated itself into many aspects of our life. Airports could not handle even a tiny fraction of their current traffic were it not for radar, and collisions would be far more frequent as radar beams guide aeroplanes on to the runway and marshal the stacking of planes waiting to land. Ships too use radar scanning to avoid collisions. Fishing boats have a 'fish finder' with a sonar scanner that gives a side view of fish in the vicinity. It enables the fisherman to guide the net around a shoal. Fish in the sea have as much chance of escaping the net as a tree has of dodging the woodcutter's axe.

The essential element in your microwave oven is an electronic tube that supplies energy. It is a magnetron like those that generate radar scans in aeroplanes. In the early 1940s Percy Spencer came to Britain to see the newly invented magnetron and was allowed to take one home with him to America. He improved the device and within a few years the Raytheon Company, for which he worked, was manufacturing eighty per cent of all magnetrons. One day he was standing beside a working magnetron and felt something hot and wet running down his trouser leg. It was a chocolate bar in his pocket that had magically changed into a chocolate drink. He placed a bag of popcorn in front of the magnetron and it promptly exploded; a raw egg did the same. The microwaves emitted by the magnetron had agitated the water molecules inside the food 2.5 billion times a second, and not surprisingly heated it up. Spencer patented the first microwave oven in 1950. It was over 6 ft (1.8 m) tall and cost $3,000.

The daily weather map on the television is the output of radar scans from satellites. Airliners scan the weather up to 200 miles (320 km) ahead so that they can skirt around storms. Watson-Watt first experimented with radar beams when working as a meteorologist, using them as detectors of snow and rain clouds. Years later when he was Sir Robert Watson-Watt, he was late driving to give a lecture on radar when he was caught by a radar speed trap.

CAPTURED IN TIME

'What is all knowledge too but recorded experience'
– Thomas Carlyle

A t one time we could send radio waves across the world, but we couldn't store them. For centuries information was lost for want of means to record it. The earliest records were etched in stone or clay that was then baked hard. These tablets are far more durable than any other form of data storage, but somewhat bulky. The Hittites had a large library of shelved stone documents meticulously catalogued, which survive to this day, but browsing requires heavy lifting.

The written word

The load for librarians and readers was lightened when parchment, papyrus and wood-pulp paper replaced rock. Reeds and quills were dipped in oak-gall ink to write, but it was slow work. A copy of the New Testament took six months to complete and a single mistake could mean an entire page had to be rewritten. The Catholic Church set battalions of monks scratching away in scriptoriums to copy ecclesiastical documents. Less pious men saw a business opportunity and their commercial copying thrived. In the fifteenth century Vespasiano da Bisticci employed fifty scribes in his copy shop.

Copying by hand was expensive, so books were only for the rich. It was so labour-intensive that even libraries had a famine of volumes and many were shackled to the shelves to discourage 'borrowing'. Around six hundred years before Da Bisticci's time a Chinese alchemist named Pi Sheng made small clay tiles, each with a common word modelled in mirror image. He could arrange these in a rack, ink the embossed words and press them on to parchment. Unusual words were made as required. Since Mandarin Chinese has thousands of characters, only the simplest of text could be printed.

European alphabets are more amenable and in 1438 Johannes Gutenberg, a German goldsmith, cast individual letters in metal. They were all the same height, so they could readily be assembled into sentences on a frame the size of the page, although this might take an entire day. Gutenberg's great invention was the printing press, which would be used with only minor amendments for the next five hundred years. It pressed a sheet of paper on to inked letters and numbers, arranged to form words and figures, and within a minute the page was completed and another sheet was being pressed. It was the first example of mass production.

The printer would scan each page to check for mistakes before printing a hundred or so copies. Even so, errors slipped through. In the reprint of the Bible in 1611 one of the commandments was: 'Thou *shalt* commit adultery', which came as a relief to many readers.

Within two decades there were printing presses in almost every European country, fifty in Italy alone. The price of books plummeted and they were no longer just for the elite. Gutenberg had democratised knowledge and given the population an incentive to learn to read. It also gave the Church more things to ban.

When, around the year 340, the Codex Sinaiticus, the oldest surviving copy of the New Testament, was copied to be made into a book, there were twenty-three thousand corrections, about thirty per page of the parchment document. Although a few were to rectify errors the vast majority were revisions to the text required by the clergy.

Good vibrations

In the nineteenth century inventors wondered whether sound could be stored just as words were contained in a book. In 1857 Léon de Martinville traced the vibrations of sound waves on to a revolving cylinder coated in charcoal dust, but it couldn't play back the sound. In 1877 his compatriot Charles Cros suggested resting a needle on a revolving cylinder and scratching a groove that varied in depth according to the volume of the sound. A second needle attached to a vibratable diaphragm could follow the groove and reproduce the recorded sound. Cros was absolutely right, but he never built the machine.

Coincidently, that same year Thomas Edison came up with the same idea to store Morse-code dots and dashes as indentations on paper strips soaked in paraffin wax. He did a rough sketch for his mechanic, John Kruesi, with the demand: 'Make this.' Within four months he had a working model which was promptly patented. He shouted into a cone-shaped horn to concentrate the sound: 'Mary had a little lamb . . .' and when it played back even Kruesi was astonished to hear the device repeat the rhyme in a scratchy whisper.

At the base of the horn was a telephone diaphragm with a stylus to gouge a groove in tinfoil wrapped around a rotating cylinder. The cylinder turned on a screw thread that moved it sideways so that the stylus cut a spiral groove. The vibrating stylus made tiny hills and dales in the grooves, thus storing the sound. A second, blunter stylus tracked the groove and converted the irregularities back into sound emitted from a different horn.

Edison called his device the Phonograph or Speaking Machine because he assumed its main purpose would be to dictate letters for typing. He listed ten likely uses, but did not include recording music. It soon became clear that this was *the* most commercial use. His record players were operated by turning a handle to rotate the cylinder and there was a listening horn to amplify the sound. Over time the horn became larger and ever more ostentatious, like a giant seashell or an enormous exploding nasturtium. There was no volume control and if it was too loud for the delicate Victorian ear something had to be done. This was probably the origin of the expression, 'Put a sock in it.'

The phonograph's reproduction of the human voice was not perfect: even the most mellifluous sounded like a distressed Dalek. The only instruments that came out well were the mandolin and the banjo. To record a pianist a horn resembling an enormous ear trumpet had to be suspended above the strings. To record the whole orchestra a gigantic horn was used.

The tinfoil on the cylinders, which only played for two or three minutes, wore out after three plays. Edison rapidly adopted wax cylinders made from candle tallow. Rivals undercut his price: 'The Oxford Cylinder Talking Machine – with a flexible horn only $14.95 equal in tone and volume of any $30 machine ever sold.' Not being musical instruments they were sold by catalogue or in bicycle shops.

A more serious competitor was on the horizon. In 1897 Emil Berliner, a German working in the United States, invented the Gramophone, which had a horizontal turntable that played flat discs the size of a modern CD. His other innovation was a stylus that didn't dig 'hills and dales': instead it vibrated sideways, leaving wiggles on the sides of the groove. Unlike the cylinders, a flat metal disc could be etched as a master from which any number of rubber records could be stamped. A variation on this was a children's phonograph with edible discs made of 'squeaking sweetmeat' which, when warmed and pressed on to a record, gave 'a true reproduction'.

During elections in England some candidates recorded their speech several times and sent out phonographs on carts to the far corners of their constituency that they were too lazy to visit. With cylinders an artist had to sing a song into the recorder dozens of times because every recording was a one-off. Famous musicians were not tempted by such marathons. Berliner's mass-produced discs changed all this and when the great Enrico Caruso sold more than a million copies every musician wanted to record.

In 1892 sound quality improved considerably when rubber discs were superseded by records made of shellac, a resin from insects. The discs were stamped out in vast numbers and were hard enough to withstand numerous plays. As a boy I was given a batch of shellac records that had been imprisoned and overworked in a jukebox. They had endured so many plays that the black shellac had worn away to grey and when I held one up to the light I could see light through the grooves. Even so they still played. They also made excellent Frisbees and if warmed, could be moulded into wavy-edged, wobbly bowls that nobody wanted.

In 1931 the RCA Victor company produced the first twelve-inch (25 cm) long-playing record to accommodate Beethoven's Fifth Symphony, but it wouldn't play on any machine available to the public. The idea was revived by Peter Goldmark at Columbia in 1948. The vinyl plastic discs had 'microgrooves' that were closer together and turned at only 33⅓ revolutions per minute, giving a play length of over twenty minutes on each side. The fifty-year dominance of 78-rpm records was at an end.

In 1933 EMI's Alan Blumlein experimented with recording separate sounds for right and left speakers, with the sounds carried on opposite walls of the grooves. But stereophonic records, using his method, didn't appear on the market until twenty-five years later.

As early as 1895 Valdemar Poulsen, a Dane, who worked for the telephone company in Copenhagen, had patented an entirely different method for recording sound. His Telegraphone was a magnetic system that stored sound on piano wires to be used as an answering machine. The sound quality was not good, so in his patent he stated that a better system would be to use a strip of insulating material coated with metallic dust that could be magnetised. He had described the tape recorder, which would not be fully realised until 1933 when the German firm AEG marketed the Magnetophone, which used plastic tape coated with iron-oxide particles.

Until this time radio broadcasts including plays were transmitted live. The BBC saw recording as a means of copying a programme for transmission to countries in different time zones. They went for a metal tape system which terrified the technicians. It was razor-sharp and ran so fast that if it snapped, it might lash around the room like a flying guillotine. In the USA, thanks to Bing Crosby's insistence on pre-recording his radio show, taping became standard practice in radio and recording studios. The great advantage of taping was that material could be edited and erased, which is no bad thing for some modern music.

The Dutch firm Philips introduced mini-cassettes containing narrower tape in 1961. They took the unusual step of allowing other companies to exploit their patent to ensure that the system would become standard. Even as late as 1967 Sony's most 'compact' tape recorder weighed a mighty 18 pounds (8 kg). Twelve years later they produced the tiny Walkman with earphones, which gave rise to the pedestrian zombie and oblivious jogger. Cassette tapes are making a surprise comeback. For huge data sets ultra-dense tapes are replacing hard drives as they can store information at much higher densities while using less energy.

In 1985 Philips in collaboration with Sony marketed the first compact discs, groove-less plastic discs read by a laser beam. They were demonstrated at the Salzburg Music Festival. Herbert von Karajan, illustrious conductor of the Berlin Philharmonic Orchestra, was bowled over and declared that, compared with CDs, everything else was gaslight. The marketed disc ran for sixty minutes but von Karajan persuaded the manufacturers to extend the duration to seventy-four minutes to record Beethoven's Ninth Symphony.

With an entirely new format the manufacturer is up against customers' reluctance to buy new equipment. Philips and Sony cannily produced 100 recordings on CDs in time for the launch. Sales of CDs rapidly surpassed those of vinyl records, though they have been declining since the advent of the mp3 file.

Mysterious images

Sound was relatively easy to record, but pictures took more time and a lot more chemistry. On a promontory not far from where I live on the Isle of Man there is a camera obscura (dark room) with eight small windows in the roof. Each window has an inclined mirror that, one window at a time, can throw a view of the outside landscape on a white table. The brilliant image in a darkened room is stunning.

Renaissance painters used a small hole in the wall to cast an image on to the facing wall. Portable camerae obscurae fitted with a lens that can be focused on a ground-glass screen at the other end of the box enable the artist to trace the scene. Unfortunately, light passing through a small hole has an annoying habit of turning the image on its head. Ninth-century Arab astronomers used the camera obscura to observe sunspots and eclipses. During a solar eclipse one of the trees in my garden left a myriad of pinhole gaps between the leaves and projected into its own shadow hundreds of small suns and from every one the moon had taken a bite. For a while the tree was a camera, but as it couldn't store the images, I captured them with my camera.

Photography was the chemistry of light-sensitive chemicals. The first-ever photograph harnessed an unlikely chemical.

In 1816 French chemist Nicéphore Niépce spread an asphalt varnish thinned with lavender oil on to a pewter plate and exposed it to the sun for eight hours. He then washed off the soft asphalt, leaving the hardened compound, which showed the view from his study window. His Heliograph (sun-writing) is dark and grainy but you can discern rooftops and a distant tree. He had recorded a moment and stored it for posterity.

Niépce joined up with another Frenchman, Jacques Daguerre, a theatre scenery painter, to explore the possibilities of the pinhole camera. Across the Channel, in the seventeenth century Robert Boyle had observed that sunlight caused silver chloride to turn black. Daguerre rubbed silver salts and other chemicals on to a copper plate, washed it with acid, exposed them to iodine vapour and then to light in a pinhole camera. But no image appeared, so he put the plate into a cupboard intending to clean and reuse it. Several days later he was surprised to see an image had emerged. The fumes from mercury spilt from a broken thermometer had developed the picture. It began to fade and he tried everything within reach to prevent it, and in those days even a poisoner's cabinet would hardly have contained a greater cocktail of noxious compounds. In the end common salt did the trick. In 1839 Daguerre patented his method for 'obtaining spontaneous reproduction of images received in the focus of the camera obscura'.

His 'daguerreotypes' were a sensation. In his own words: 'I have seized the light; I have arrested its flight.' The sharpness and detail of his pictures was astonishing. American poet and physician Oliver Wendell Holmes, a keen amateur photographer, called them 'Mirrors with a memory'. The exposure time was three to fifteen minutes, so the subject had to be stationary. Ninety per cent of daguerreotypes were formal portraits, and the sitter's head was clamped in a steel brace. Its grip probably accounts for their strained expression. Abraham Lincoln was photographed by Matthew Brady, who chronicled the American Civil War with his unflinching camera lens. Lincoln's majestic portrait was widely distributed and he said: 'Make no mistake, gentlemen, Brady made me President.' Alfred Lord Tennyson took the opposite view and berated a photographer: 'I can't be anonymous by reason of your confounded photographs' – the first paparazzo had annoyed a celebrity. Other notables, such as Victor Hugo and Queen Victoria, were enthusiastic amateur photographers and Lewis Carroll took lots of snaps of young girls. Daguerre sold his patent to the French government for an annual pension of six thousand francs (around £230,000 today).

The greatest drawback of the daguerreotype was that the photographs were one-offs and couldn't be copied. However, in England William Henry Fox Talbot, a scientist and country gent, had independently discovered the invisible latent image and realised that he could use much shorter exposures to light and develop the image later in his darkroom. He also experimented with paper impregnated with silver chloride to make negative shadow pictures by placing objects such as a lace doily on top of the sensitised paper. The result of exposing it to light was a white image of the doily on a black background.

Fox Talbot produced these small negatives in wooden boxes that his wife described as 'mousetraps'. It was not long before he managed to produce an unlimited number of positives from a single negative. This technique was the future of photography. Fox Talbot's astronomer friend Sir John Herschel was the first to coin the word 'photograph' (light-writing) and also 'negative' and 'positive' photographic images. Fox Talbot also took the first-ever flash photograph, in 1851.

Transparent glass negatives would produce far better positives than paper did, but glass couldn't absorb the chemicals that captured the image. Egg white and snail slime were used to make silver grains stick to glass but neither were satisfactory, much to the relief of hens and snails. Another Englishman, Frederick Archer, tried collodion, a mixture of gun cotton, ether and alcohol which formed a thin, clear film designed to cover wounds. Archer dissolved light-sensitive silver salts into collodion together with a developer that could convert them into silver. After exposure to light it developed spontaneously while still wet inside the camera. In bright light the exposure was instantaneous. I have omitted much of the messy preparatory chemistry before the plate got into the camera.

Long exposure times meant bulky tripods to keep the large cameras steady. Roger Fenton, an English photographer who documented the Crimean War, travelled around in a horse-drawn pantechnicon containing cameras, massive tripods, boxes of heavy glass plates, chests of chemicals, Winchesters of distilled water and a heavy-duty tent that served as a darkroom. In the Crimean summer the heat inside the tent was unbearable, unless the incessant wind blew it down.

In 1871 another Englishman, Dr Richard Maddox, replaced collodion with gelatin, a gel that dried hard but swelled if wetted, thus allowing the developer chemicals to reach the silver. His 'dry-plate' method was far better than trying to handle wet negatives, so it became the norm, as did Fox Talbot's silver-bromide printing paper.

Smaller, hand-held cameras were now available and some were simple to use. Flinders Petrie, the father of Egyptology, took surprisingly clear pictures of artefacts while excavating in the desert using a pinhole camera that had previously been a biscuit tin. But cameras were now acquiring viewfinders to frame the shot, and controls for changing the shutter speed and adjusting the aperture, which together determine the exposure: the amount of light that reaches the film. Small cameras meant less conspicuous photographers. The public were accustomed to seeing pictures of high and mighty people, but now the camera's eye captured the life of the urban poor: flower sellers, chimney sweeps and street urchins in the slums. Seeing these images, some people were aroused to do something to relieve their condition.

Someone said, 'The camera never lies,' but sometimes it did. Dr Barnardo took 'before and after' photographs of his rescued children, but he wasn't above dirtying up the 'before' boy and ripping his clothes a little more to increase the contrast with the spruced-up 'after' young gent. It was a small deception for the greater good.

Meanwhile the middle class rushed to studios to have family portraits taken and at the seaside trippers were snapped in the open air by photographers who produced their photo on the spot with a rubber bag of developer dangling beneath the tripod. Then George Eastman turned the world into happy snappers. Eastman was an amateur photographer who founded the Dry Plate Company in New York. He released his Number 1 Kodak camera in 1888. It was an unimpressive box until you looked inside. Instead of a glass plate there was a roll of paper covered with gelatin and bromide. The paper ran on spools that needed to be wound on before the next shot was taken. Eastman's improved box camera used flexible nitro-cellulose film. Celluloid had previously been used to make detachable shirt collars, ping-pong balls and false teeth. Its only fault was a tendency to spontaneously burst into flames, which made for exciting dentures.

The camera took 100 photographs and, with the exposed film still inside, it was posted to the factory to be developed. It was returned loaded with a fresh film. The firm's motto was: 'You press the button; we do the rest.' Next came the Kodak Box Brownie. It was designed for children and got its name from the impish 'Brownies' in a children's book. By the end of the nineteenth century the Kodak brand was universally known, although the word meant nothing at all. Eastman's favourite letter was K: 'It seems such a strong, incisive sort of letter.' So he juggled letters to invent a new word beginning and ending with K. Curiously, he hated having his picture taken.

By 1905 there were four million amateur photographers in Britain alone and Kodak cameras led the field. The company's laboratories produced numerous innovations, some by accident. A technician studying the light-refracting properties of various materials put a sample into the machine that refused to come out. He had ruined an expensive piece of kit and invented superglue.

The first colour photographs were taken independently in Britain by Thomas Sutton and James Clerk Maxwell in 1861. They and others couldn't find a proper colour balance. The Lumière brothers in France produced Autochromes in 1907 using starch grains dyed in primary colours that blocked some colours of light while letting others through.

The soft tones of Autochromes are charming, but not true to life.

In 1935 two New York music students, Leo Godowsky and Leopold Mannes, experimented with coloured film in their kitchen. They were recruited by Kodak and the result was Kodachrome. The film had three layers of emulsion each sensitive to one of the three primary colours used in colour printing – cyan, magenta and yellow – and mixing them could replicate almost every colour in nature. It was tricky to process a finished roll, so it had to be sent to the factory in its light-proof cassette. Kodachrome was marketed in 1936 as transparencies that had to be seen with a hand-held viewer or projected as 'lantern slides'. Six years later colour prints were available.

Eastman gave away his substantial wealth to worthy institutions. By 1976 Kodak had ninety per cent of film sales and eighty-five per cent of camera sales. The company had invented the digital camera the previous year. It weighed in at 9 lb (4 kg), took twenty-three seconds to take a black and white image and stored its images on a cassette which had to be transferred to a custom-made device that sent them to a television set. Kodak's directors couldn't grasp why anyone would want to view their photos on a TV screen. Steven Sasson, the digital camera's inventor, stressed that it was a prototype, but it had the potential to change how pictures would be taken in the future. They still didn't get it. Kodak underestimated the rise of digital photography and in 2012 the corporation filed for protection from bankruptcy.

The digital future was unimaginable. Who would have thought that in 2013 researchers in Illinois would invent a soluble camera. Yes, one that dissolves. It was marketed as a spy's gadget.

The magic box

Motion pictures are an illusion. They are a series of still photographs flashing by in sequence. The principle was neatly demonstrated in 1826 by an Englishman, Dr J.A. Paris, who painted a canary on one side of a cardboard disc and an empty cage on the other. When he spun the disc the bird was magically imprisoned in the cage.

When I was a lad seaside amusement arcades still had coin-operated mutoscopes, better known as peep shows. A popular title was 'What the butler saw', but the pictures ran out before the butler (or the viewer) saw anything worthwhile. Mutoscopes were free-standing tin boxes containing a drum of 'flicker photos' which, when you turned the handle, rotated to briefly reveal each picture in turn giving the impression of a moving image. I recall drawing a matchstick man on the corner of every page of a book. When I riffled the pages the man danced and waved his arms. The first cartoon film was drawn by an Englishman, Stuart Blackton, in 1900 and until the advent of digitisation all cartoon films were made this way.

So the principle of creating apparent motion was well established, but what was lacking was a camera that could take a rapid series of exposures of moving objects. This was invented in 1878 by Eadweard Muybridge, a man determined to make a name for himself. Indeed 'Eadweard Muybridge' was an invented name and he also assumed the title of professor. He left Britain for the United States to seek his fortune selling books. This was the slow lane to riches, so he took up photography and he was good at it. Calling himself Helios, he took stunning landscapes of virgin America. He boasted that he would 'cut down dozens of trees to acquire a picture on which he had set his heart'. His stereoscopic photographs sold in their hundreds of thousands.

Muybridge described himself as a man who would 'stand no trifling'. His flirtatious wife was having an affair, so he confronted the man and expressed his displeasure by shooting him through the heart. At the trial he admitted that his only regret was that his victim 'died too quickly'. A jury of husbands cheered when the judge announced a verdict of 'justifiable homicide'.

Leyland Stanford, a railway magnate and benefactor of Stanford University, sponsored Muybridge to show whether the legs of a galloping horse were ever all off the ground at the same time. Muybridge managed to capture the motion of a stallion with tripping wires that triggered a row of twenty-four cameras. The shutter speed that 'froze' every stage of the horse's stride was an unprecedented one-thousandth of a second and it proved that all the hooves were indeed off the ground simultaneously.

Muybridge became obsessed with capturing the motion second by second. The subject was naked men running, turning somersaults and wrestling. He also froze the stream of water thrown from a bucket by a young woman. Why she was naked only Muybridge knows. He also photographed himself naked swinging an axe and surrounded by a circle of cameras that triggered simultaneously to catch his swing from all angles.

His next step was to print a short series of photographs

and place them in a rotating cylinder very like the Zoetrope, a children's toy with drawings pasted on the inside of a spinning drum and viewed through a series of slits that served to animate the pictures. At the Chicago World Exposition of 1893 Muybridge managed to project images from his Zoöpraxiscope on to a screen 10 ft (3 m) tall. Although each showing lasted only a few seconds, the audience was astounded.

In France Étienne-Jules Marey invented the multi-shot photographic 'gun', complete with rifle butt. When he pulled the trigger it rapidly exposed a dozen photos on a circular plate that could then be spun in a projector to reveal a bird in flight. Later he was able to shoot 120 consecutive pictures on a strip of paper, each with an exposure of one-thousandth of a second. Edison was aware of Marey's advances and when Eastman's celluloid film became available, he used a continuous loop of film in his improved peep-show machine, the Kinetoscope of 1891. It was essentially a movie for one person at a time. The film loop lasted only twenty seconds, so the 'action' couldn't dawdle. One showed Annie Oakley getting her gun, another a man sneezing. To save money, Edison cut Eastman's 70-mm-wide film down the middle and created 35-mm film with sprocket holes at each side to advance the film inside the camera. He had created the movie film stock of the future.

The cine camera was invented in 1886 by a French artist, Augustin Le Prince. He built working models of a camera and a projector using celluloid film. Remnants of his footage that survived were taken when he was in England and include a tour of his father-in-law's garden and rush-hour traffic in Leeds. He never got to exploit his inventions. On a trip to demonstrate his equipment he boarded a train in Dijon bound for Paris, but he never arrived and no trace of him or his equipment was ever found.

At the same time, back in England William Friese-Greene, a portrait photographer, dedicated his life to cinematography. His early experiments used sensitised, oiled paper which proved that his camera was ideal for ripping paper. By 1890 he had developed and printed on celluloid scenes of hansom cabs passing Hyde Park Corner.

A movie camera just had to take pictures rapidly. The projector was a more difficult challenge. If the film whizzes through the projector continuously the image is just a blur. For every frame to register, the film has to continually hesitate for an instant behind the lens as the shutter opens. The paradox of moving pictures is that to make the motion flow visibly it has to freeze. The mind's eye retains each image long enough to ensure a smooth transition.

In 1895 Auguste and Louis Lumière patented a movie camera and projector. They used the mechanism from a sewing machine to drive the hesitant advance of the film. There was a viewing for paying customers in the basement of the Grand Café in Paris. The Lumières projected ten short films and the audience sitting in the intimate darkness were amazed. They saw workers streaming out to the two brothers' workshop, a blacksmith hammering a horseshoe on his anvil, and when a train steamed into a station they dived for cover under their seats. The magic of the cinema had been born. The press were ecstatic: 'It is life itself.' At the Paris Exposition of 1900 the Lumières erected a huge screen. It was estimated that 1.5 million people saw the programme over the duration of the exhibition. They called their equipment the Cinématographe and the name stuck.

Edison woke up to the potential of motion pictures and bought the rights to a camera/projector. He also blocked and bought out rivals in the USA and enforced a monopoly of patents until 1917. In California he established the first custom-built film studio, known as the Black Maria. It was designed and built by a Scot, William Dickson. There was a single fixed camera facing a stage. To exploit natural light the roof folded back and the building was on a circular track so that it could follow the sun.

There were no cinemas (Greek for 'motion') to begin with, so films were shown in music halls and unused buildings until Edison built sheds called Nickelodeons because the entrance fee was five cents. In New York he tried to monopolise the picture houses and even sent in goons to rough up his rivals. Many of the patrons were immigrants who had little or no English, so silent films were ideal. The sheds established movies as cheap entertainment for the working class. Refined people went to the theatre instead.

The movies began to tell stories and to develop the techniques for the task. In Brighton on the south coast of England George Smith shot *The Kiss in the Tunnel* and invented the close-up in the late 1890s. In 1907 in America David W. Griffith, an unsuccessful playwright, began acting in short movies. One day the following year the director didn't turn up to film him wrestling a stuffed eagle, so Griffith, who knew nothing about photography, moved behind the camera and directed the film. After making a hundred or so 'shorts' he turned his hand to epics. Later D.W. Griffith would never tire of reminding people, in his modest way: 'How small the world was before I came along. I brought it to life: I moved the whole world onto a twenty-foot [6 m] screen.' He *was* a gifted director but the techniques that he is often said to have developed such as soft focus, the flashback, fade-out and dissolve were invented by his cameraman, Billy Blitzer. Griffith did get the camera out of the studio and into a balloon for extreme long shots and on top of a speeding car to keep up with galloping horses. However, it was a Russian, Sergei Eisenstein, who invented the visual language of motion pictures with his mastery of montage – the art of editing and assembling scenes into a film.

The film industry thrived in the sunlight of California, although the very first feature-length movie was an Australian effort, *The Story of the Kelly Gang,* in 1906. American film-makers converged on the real-estate development of Hollywoodland, so called because there wasn't a holly tree to be seen. Despite being the home of sagebrush, it became the world centre of romance and glamour, as well as high drama and hedonism – some of it on the screen. At first actors were not celebrities. A hotel sign read: 'NO ACTORS OR DOGS.' Even the most popular actors were anonymous because the producers feared that if they became known they might demand a rise. The first movie star, Florence Lawrence, 'The Biograph Girl', earned $13,000, a fortune in those days. But like all stars she faded, so she swallowed ant powder to exit left. Hollywood became a suicide hot spot with failed wannabes and has-beens vying to devise the most bizarre fade-out.

In the 1920s one of the biggest stars was Rin Tin Tin, a German shepherd dog found in an abandoned dugout on a battlefield in France. He was described as 'a sensitive actor'. Film audiences suspended their disbelief to such an extent that the dog received two thousand fan letters a week. Some voiced their disappointment that he didn't reply. A decade later Mickey Mouse received two thousand letters a day, more than any human actor. A Rough Collie called Lassie was such a good actor that no one noticed she was a he. The MGM lion was also a celebrated actor. As the lion was being filmed, Samuel Goldwyn insisted that he should roar so loud that it would be heard even in silent films.

It became customary to accompany silent films with music from a piano or violin in the auditorium. At first they played popular tunes of the day unrelated to what was happening on the screen. As the heroine sobbed over the grave of her husband, the piano played jolly jingles, perhaps to cheer up the mourners. Later the musicians were requested to watch the picture and improvise a suitable accompaniment. Griffith's *Birth of a Nation* (1915) warranted a thirty-five-piece orchestra playing Wagner's *Ride of the Valkyries*. Eisenstein's masterpiece *Battleship Potemkin* (1925) had an original score written for it. In several countries the film was screened, but the music was banned.

In 1927 everything changed. Warner Brothers' studio was in financial trouble and gambled everything on their next picture, *The Jazz Singer*. The film wasn't that good, but when its star, Al Jolson, said, 'You ain't seen nothing yet,' he wasn't kidding. Movies had found their voice. Irving Thalberg, Head of Production at MGM, declared it was 'just a passing fancy', but thirty-five finished silent pictures ready for release were now obsolete.

Crowds flocked to *The Jazz Singer* and the other studios tried to save their expensive movies by adding sound. In Britain Alfred Hitchcock released both silent and sound versions of his thriller *Blackmail*, because many cinemas were not yet wired for sound. A film critic thought the silent version was far superior because talkies were 'a poor conjunction of two dramatic forms'. The paying public, untroubled by conjunctions, did not agree.

Since the 1890s inventors had been trying, with varying degrees of success, to mate film with sound from the gramophone. It was difficult to ensure *exact* synchrony with an actor's voice and even the slightest deviation was useless. Frenchman Eugène Lauste took a different approach and in 1906 patented a method for photographically printing sound vibrations on to film. This could be read by a photocell and amplified as sound. As long as the soundtrack matched the action on the adjacent frame all was well. The clapper board was the starting 'pistol'. As the filming of a scene began the clapper made a clear sound signal on the sound track. Thus sound and image could be easily matched. The first commercial performance of a feature film with a sound track imprinted on the film was *Der Brandstifter* (The Arsonist) in 1922. By the early 1930s all films were 'talkies'.

The coming of sound was a blow for many stars. As one critic put it: 'If I were an actor with a squeaky voice, I would worry.' Some of the biggest stars of the day retired prematurely. This era with microphones secreted in flower bowls or disguised as telephones was wonderfully parodied in the musical *Singing in the Rain*.

The next development was colour. Until colour film was available some short films had every frame tinted by hand. In 2012 an archive of the work of Englishman Edward R. Turner was discovered. It contained lovely colour shots of his children, a parrot and a goldfish in a bowl. The pictures dated from 1901–2. Four years later George Smith produced his two-colour Kinecolor. The original Technicolor was also a two-colour film and we had to wait until 1930 for movies in full colour.

By 1948 cinemas in the USA had an audience of ninety million patrons a week. Within just three years total attendance plunged to fifty-one million. Film magnates had underestimated the impact of television. To counter the slump studios turned to bigger screens. Wide screens were not new. In 1927 French director Abel Gance's silent epic *Napoléon* required three screens and three projectors. Seven years later stereoscopic sound was added to this film. In 1952 Twentieth Century Fox adopted a different filming system that had been known since the 1920s. The camera lens squeezed the image laterally so that it fitted on to standard 35-mm film. The projector lens then stretched it out again on the Cinemascope screen.

Some studios gambled on 3D instead. Short films in 3D had been produced in the 1930s, but the first feature-length 3D picture was *Bwana Devil* in 1952. To exploit the film's only asset lots of spears were thrown at the audience and the advertising slogan was: 'A lion in your lap and a lover in your arms.' How one copes with both at the same time, goodness knows. One executive was reputed to have said that so many things would be thrown at the audience, they would start throwing things back – yes, probably rotten tomatoes. The critic's assessment of the film was: 'Inept.'

Today more than twice as many films are made in India than in Hollywood, which echoes with empty sound stages. They are again turning to 3D for a lifeline. Perhaps instead they could try putting as much effort into their screenplays as their explosions and car chases. While we are on the subject,

I am happy to report that in 2013 the BBC stopped producing 3D television programmes because of their waning popularity.

Cinemas were not just pleasure domes, for even in the early days they examined the world around them. In England James Williamson re-enacted world events. He filmed the attack on the British Mission during the Boxer rebellion of 1899–1901 in China. The actors were family and friends and the Mission was his house. By the late 1890s cinema-going patrons had seen real-life footage of events such as the Epsom Derby, Buffalo Bill's Wild West Show and the naval review to celebrate Queen Victoria's Diamond Jubilee.

Before television made them redundant Movietone or Pathé newsreels were always screened before the feature film. They delivered a pictorial presentation of what was going on in the world. And no matter how distant or exclusive the event, the man in the street had an invitation for the price of a cinema ticket.

THE CAT'S WHISKERS

'The wireless music box has
no imaginable commercial value'
– A broker advising in the 1920s against investing in radio

For all its merits wireless telegraphy could only transmit Morse code, which just required turning the current on and off. Carrying speech and other sounds on a radio wave would be a challenge. The signal had to change the shape of the 'carrier' wave – that is, modulate the signal. At the receiving end the carrier wave had to be stripped away and the signal converted back into sound. But what could be the receiver?

The simplest answer was a crystal set. As early as 1880 Frenchmen Pierre and Jacques Curie (Marie Curie's husband and his brother) discovered that some crystals became electrified by pressure. The current was very weak but if a fine wire, which became known as the 'cat's whisker', was poked into the crystal the electrical current flowed only one way.

In the early days of broadcasting the crystal set harnessed the current from a crystal of galena (an ore of lead) to become the standard radio receiver. Incoming radio waves received by an aerial created only a tiny electrical signal, so the listener had to use earphones. What was needed was an amplifier.

Valves to the rescue

The answer came from an Englishman, John Fleming, who had been Marconi's right-hand man. He recalled that when searching for the ideal light bulb Thomas Edison had made an interesting observation. While testing an electronic tube in a vacuum he found that a heated wire emitted a one-way current. In 1904 Fleming invented the thermionic valve, later called the vacuum tube. It was a sealed glass tube containing two electrodes, one heated (hence 'thermionic'), the other cold, and the current flowed in one direction. This became the receiver in all radio sets and the crystal set was relegated to do-it-yourself kits for schoolboys to assemble.

In 1906 Austrian physicist Robert von Lieben added a third electrode which substantially amplified the impulse, but he died shortly afterwards and never received the recognition he deserved. Instead it went to Lee de Forest, an American, whose three-electrode valve, when coupled to an oscillator, gave a stronger signal and became an essential component in radio sets. De Forest denied all knowledge of the work of Edison, Fleming and von Lieben, which seems improbable. With better valves and more powerful circuits, loudspeakers were fitted into radio sets, but they should have been called 'quietspeakers' because one could only hear their output if everyone in the room kept quiet. Modern loudspeakers were not invented until the 1920s.

In the late 1940s Bell Telephone Laboratories in the United

States set out to find a replacement for the bulky and expensive valves. William Shockley, who always took sole credit for any discovery in his laboratory, outlined the principle of a much smaller and cheaper device. Two of his colleagues, Walter Brattain and John Bardeen, developed the transistor in 1948. Elements such as germanium and silicon are electrically odd. They are neither conductors nor resistors and are called semi-conductors. When three wires were poked into slices of germanium they could do everything a valve could, except amplify sound. This was rectified by putting the transistor on a chip of silicon with other components.

At first transistors were used only in hearing aids. It took a decade to realise their huge potential. Their small size made portable radios, mobile phones and satellites possible. The previous 'portable' radios were the size and weight of a packed suitcase. The silicon chip became crammed with ever smaller components and eventually printed circuits were no bigger than a full stop on this page. The future had arrived.

The pioneers of radio considered it to be a means of communication like the telephone. Few anticipated that broadcasting of entertainment would become a major function. Tesla demonstrated voice transmission by radio in 1893, although it was Marconi who was lauded as the inventor of radio when he did the same thing two years later. He did, however, patent an invention that enabled several stations to transmit on different wavelengths to avoid interference from other stations.

In 1915 a transmission from the US Navy wireless station in Virginia was received by French military on the top of the Eiffel Tower. Two years later Marconi experimented with

Very High Frequency (VHF) transmissions in Wales to a receiver 20 miles (32 km) away, although VHF did not come into general service until 1936. In 1924 Marconi spoke to Australia by short-wave radio.

Radio days

The first radio broadcast for 'entertainment' was made by a Canadian, Reginald Fessenden, in 1906. In his fun-filled programme he played Gounod's 'O Holy Night' and read from the Bible. The lucky audience was confined to wireless operators on passing ships. In 1920 Marconi began a regular news service in England that included the Englishman's favourite, the weather forecast. The programmes were picked up 1,500 miles (2,400 km) away. Dame Nellie Melba sang on the opening day. When the engineer explained that she would be broadcast from the aerial pylon she replied: 'If you think I am going to climb up there, young man, you are greatly mistaken.'

Eight months later Westinghouse instigated radio transmissions from Pittsburgh and its first broadcast covered the presidential election. In 1922 the British Broadcasting Company (Corporation from 1926) began transmitting daily from London. Within five years there were two BBC channels (called programmes in those days) and over two million radio sets to receive them.

Because the signals were weak the pioneering listeners had to have an aerial wire slung on poles 30 ft (10 m) tall in the garden. Families in England clustered round the 'wireless' for news and entertainment; in the United States it was the Radiola. The sets were large wooden or Bakelite boxes with knobs to twiddle when, as usual, it strayed off station, and the valves took almost five minutes to warm up. Yet radio soon became the main source of entertainment and it was delivered directly to your parlour at the speed of light. Indeed you could hear the music play before it reached the audience at the back of the concert hall. Radio was so compelling that in 1938 Orson Welles's dramatisation of H.G. Wells's *The War of the Worlds* panicked listeners into believing that aliens really had attacked New York and many fled in terror.

Radio was not the first system to carry entertainment into the home. In the 1890s musicians, singers and preachers could be yours on demand. French inventor Clément Ader took microphones into concert halls to transmit music live and in stereo by telephone. Provided the microphones weren't situated too near to the bass drum or trombones the transmission quality was reasonable. Ader's Théâtrophone cost a yearly fee of 180 francs (equivalent to three months' rent for an apartment in Paris) and fifteen francs per performance. Each subscriber got a headphone and a cork-lined cubicle that cut out extraneous noise. The general public could sample the entertainment in two-and-a-half-minute instalments in coin-operated booths. One Paris hotel boasted of having the Théâtrophone in each of its 200 bedrooms.

The fad crossed the Channel to England and gave clients access to eighteen London theatres and several churches. Ader's company also marketed the Electrophone table with numerous earphones for family listening. A Budapest franchise with fifteen thousand subscribers provided sports updates. As radio reception improved it replaced telephone downloads in the public's favour.

The hypnotic box in the corner

Several inventors thought that if we could transmit speech, why not pictures? It was obvious that it would be impossible to send an entire picture. The image would have to be broken down into pieces that could be transmitted and then reconstituted on arrival.

The road to television began earlier than you may think. In 1880 American inventor George R. Carey shone the image on the back of a plate camera on to an array of light-sensitive cells. Each cell received light from a different area of the image and this generated a tiny current to spark a screen made up of a matching array of light cells. Carey estimated that to get even the crudest image would require fifteen thousand connections, so he gave up. He had understood that each fragment of the image had to be transported independently and his idea was not a million miles away from the principle of a modern LCD screen.

In 1884 German scientist Paul Nipkow built a different device which he called the 'electrical telescope'. It converted images into electrical signals by means of a spinning disc with a spiral of small windows cut out of it. This broke down the image into a pattern of flashes of light. Photocells detected the light and dark areas and sent the information down a wire to a lamp that projected the sequence of light and dark through windows in an identical disc spinning in synchrony with the first. Thus the original image was reassembled on to a screen. The principle was fine, but the pictures were not.

In 1897 another German, Ferdinand Braun, invented the

cathode-ray tube. It was a sealed glass tube with all the air removed and a hot metal electrode at one end. If given sufficient voltage the electrode emitted invisible rays that shot down the tube. When Braun daubed fluorescent paint over the far end of the tube it glowed in mysterious patterns. Using magnets he focused the beams into a bright spot.

In 1906 Boris Rosing, a Russian student of Nipkow's, used a spinning disc to disassemble an image but replaced Nipkow's second disc with a cathode-ray tube. With Tesla's electromagnet coils he could make the light spot swing rapidly back and forth in parallel lines across a screen, 'depositing' the image as it went. This was a vital development but because Rosing had no means to amplify the signal it produced indistinct pictures. Before he could improve his device he was lost in the Bolshevik revolution.

Alan Swinton, a Scottish electrical engineer, believed that Rosing had used the right receiver but the wrong transmitter. The cathode-ray tube could do *both* tasks electronically without the need of a rotating disc. It could read the original image like an eye scanning a book. Swinton later published the details of his proposed machine but he never built it. Had he done so he would have been the father of television, twenty years ahead of anyone else.

In the meantime Nipkow's discs were kept spinning thanks to another Scot, John Logie Baird. Although trained as an electrical engineer Baird was a failed businessman reduced to selling shoe polish and razor blades. His career in shoe polishery was dashed by ill health and while convalescing he began to invent things such as the glass razor blade as well as the 'medicated undersock'. Then he turned his attention to what would be called television. He was broke, so he relied on a friend who sent parcels of electrical devices, including a light cell. He also commandeered useful equipment such as a tea chest, a bicycle lamp and a biscuit tin sadly bereft of biscuits.

Baird persevered with spinning discs almost as big as car wheels and managed to transmit a tiny image of a wooden cross. In 1925 he demonstrated his device at the Royal Institution and at Selfridges department store, where over three weeks a million people viewed his device. He also took it to the *Daily Express*, but the news editor's response was to get rid of this lunatic with his machine for seeing by wireless. He warned the hack he was sending to confront the dishevelled inventor: 'Watch him, he may have a razor on him.'

Baird was trying to impress both the scientific community and the general public. In the demonstration he transmitted an image of the grotesque head of a ventriloquist's dummy called 'Stookey Bill'. That same year he paid an office boy called William two shillings and sixpence to sit still and become the first person to appear on television. The camera needed very bright light and the big batteries sparked and set fire to the sitter's hair. Despite the poor quality of Baird's images, Lee de Forest, who had invented an electronic valve, dismissed the cathode-ray tube and was certain that Baird's mechanical machine would win out.

From 1925 the BBC began irregular television broadcasts using Baird's system. His first transmission included Gracie Fields, a big star in those days. Unfortunately, the picture and the sound were not in synchrony. It began with a silent Gracie miming, followed by her voice booming out of a blank screen. Irish actress Peggy O'Neil was the first person to be interviewed live on TV for her critical comments on this new medium: 'It's a rather jolly experience . . . to say the least, it's very wonderful. And what a top-hole present a Televisor would make.'

The Televisor, Baird's first TV set, went on sale in 1930 for £26, well above the means of the man in the street. It could deliver sound and pictures, but the screen was tiny and the picture even smaller. Initially the viewing family were gathering round an illuminated postcard. They watched in darkness because the set had to be plugged into the light socket. Six years later came the Ultra set, which was as big as a chest of drawers. As the cathode-ray tube was held vertically in the cabinet the picture was reflected in a mirror that lined the half-open lid. Viewers longing for a larger screen bought what looked like a Zimmer frame holding a large oblong magnifying glass in front of the screen.

In 1936 the BBC launched a regular service of advertised programmes transmitted in alternate weeks by Baird's system, and a fully electronic system was developed at EMI (Electrical and Musical Industries) in London. Within three months Baird's system was taken off the air in favour of a clearly superior one. EMI's picture was far sharper because the electronic tubes scanned far more lines on the screen than Baird's system. Baird's cameras frequently overheated and vibrated so violently they had to be bolted to the floor. In contrast, EMI's cameras were relatively lightweight and mobile.

Not a single component of Baird's design would be found in subsequent TV equipment. He had backed the wrong horse. His importance was that he alerted the public to the possibilities of television, not just through his programming, but also his demonstrations of colour and stereoscopic (3D) TV in 1925 and transmissions across the Atlantic three years later.

In 1923, the year that Baird patented his system, a Russian émigré named Vladimir Zworykin filed a patent for his cathode-ray transmitting tube in the United States. The following year he patented his cathode-ray receiving tube. Although he had a complete television system, companies were not interested. Perhaps they were influenced by De Forest's proclamation that even if television were technically feasible, it would be a commercial flop.

First Westinghouse and then RCA (Radio Corporation of America) supported Zworykin, but his demonstration in 1924 displayed pictures inferior to Baird's. A director was reputed to have told Zworykin's line manager to put him on something else more profitable. Eventually RCA spent $4 million to perfect his apparatus. Then out of the blue Philo Farnsworth, a farm boy from Utah, patented a television system with a sixty-line screen. It cost RCA $1 million to buy the rights to his patent.

Another Russian émigré, Isaac Shoenberg, was independently developing an electronic television system for EMI in London and had invented the first practical TV camera. It was his technology that had ousted Baird from the BBC and enabled it to transmit at VHF (Very High Frequency), high definition at that time. The only problem was that the high-power transmissions interfered with the cameras and the cutlery in the canteen gave off sparks.

Lord Reith, the director general of the BBC from 1927 to 1938, had fixed ideas on programming. He admitted that he knew what the public wanted, but he was damned if he would give it to them. His successors just conceded that some programmes were distressingly popular. In 1930 Reith issued a guide for newsreaders which dictated that 'off' was pronounced 'orf', 'issue' was 'issyou' and 'because' was 'becourse'. He was outraged when northerner Wilfred Pickles gave 'aircraft' a hard 'a'. The unseen radio presenters had to wear full evening dress to maintain the proper demeanour.

Even those 'in the know' vastly underestimated the impact of television. The Hollywood film moguls believed people would soon tire of staring at a wooden box in the corner. The first director of BBC television thought that the viewer would rather watch London rush-hour traffic than the latest musical movie.

In post-war Britain a television set was a still a luxury item and you might occasionally visit a friend just to watch the 'telly'. Those who couldn't afford a TV set erected an aerial on the roof so no one would know. In 1952 everything changed when a million TV sets were bought to watch the coronation of Queen Elizabeth. From then on we were hooked.

In 1969 viewers in thirty-three countries saw the moon-landing live. I didn't because I was in Cork and just as Armstrong was about to step on the moon the programme switched to an interview with Collins's aunt in Wexford.

In the early days viewers were kept fit by continually rushing to adjust the picture. The most used knobs were 'horizontal hold', when the picture broke up and slipped sideways as if on an urgent errand, and 'vertical hold', which stopped the picture from tumbling. When the technicians cured these pictorial acrobatics the settee became ever more comfortable. Sofa-stuck viewers were encouraged by the invention of the remote control, the brainchild of Robert Adler in the United States. His first device, aptly named the Lazy Bones, was attached to the TV set by a cable. The first cordless remote control was the Flashmatic in 1955, which used a beam of light from a photocell to activate electronic switches in the TV set. However, almost any light could do the trick, so just switching on a table lamp could change channels or switch off the telly. Adler's Space Command used ultrasonic waves and became the sofa surfer's best friend until it was superseded by infrared remote controls.

Surveys in the United States in the 1990s revealed that by the time children reached eleven years old they had on average spent 1.7 years watching television. By the age of eighteen they had seen twenty-six thousand murders. Now nothing is unseen and we witness riots, wars, earthquakes and assassinations. Who will ever forget those planes stabbing into the Twin Towers? On the other hand, television has taken us to the depths of the ocean and on to the scorched red plains of Mars. Fearless TV cameras have probed where we could not go, such as the deadly interior of a nuclear reactor and my innards both top and bottom.

People who watch six hours of television every day can on average expect to die five years earlier than those who don't watch any. It is curious that we now spend so much of our lives watching TV while at the same time we are being watched by TV. The proliferation of closed-circuit TV echoes George Orwell's vision of a population under constant surveillance.

Chronic viewers were often warned against getting 'square eyes'. Aptly St Clare of Assisi is the Patron Saint of television *and* eye diseases. The word 'television' is half Greek and half Latin – no good will come of it.

ALL THINGS GREAT AND SMALL

'The telescope is a device . . . enabling distant objects
to plague us with a multitude of needless details'
– Ambrose Bierce

Television and cinema created stars and opened our eyes to fantasy worlds, but for centuries scientists had been searching for unknown worlds that were unimaginable. Over a thousand years ago an astronomical observatory was established in Baghdad from which astronomers gazed at stars in the night sky. Great advances were made by the ancient Greeks. Aristarchus of Samos suggested that the earth and the planets circled around the sun, and the earth was spinning around, giving us day and night as the globe turned to face the sun and then returned into shadow. Unfortunately, an entirely different interpretation of the solar system from an unreliable source was accepted for well over 1,500 years.

In the second century Ptolemy, an influential Greek mathematician, geographer and astronomer, announced that the earth held pride of place at the centre of the universe, with all other heavenly bodies orbiting around it in perfect circles like sycophantic courtiers. This was exactly what the Church had always known.

Even with the naked eye astronomers could observe the

phases of the moon, the daily passage of the sun, the constellations of stars and the occasionally chilling eclipses. The ancient Greeks mapped and named the constellations and applied their findings to everyday life, as the poet Hesiod revealed: 'When the Pleiades rise it is time to use the sickle, but the plough when they are setting . . . when Arcturus ascends from the sea and remains visible for the entire night, grapes must be pruned; but when Orion and Sirius come in the heaven . . . the grapes must be picked.'

A mechanical marvel

The astronomical knowledge of the ancient Greeks was outstanding and we know this because of an almost unbelievable find. In 1900 a sponge diver in a copper helmet and lead-soled boots was trudging across the sea floor 180 ft (55 m) down, off the island of Antikythera in the Aegean Sea. Amid the gloom he was confronted with a horrendous sight: 'A heap of dead naked women, rotting and syphilitic . . . horses . . . green corpses.'

The chief diver went down to investigate and found the remains of an ancient Roman ship overloaded with bronze and marble statues. They were Greek treasures 'acquired' by a wealthy Roman, probably on their way to decorate his villa in the capital. They had rested undisturbed on the bottom of the sea for more than two thousand years and had become embedded in a natural calcareous cement.

The divers retrieved all the relics, jewels and a dozen wonderful life-sized statues. Not surprisingly, a lump of cement with a small wheel in it was neglected. It was not until 1958 that a British physicist examined what was inside.

He discovered small, beautifully engineered bronze cogged wheels like those in a clock. Subsequent examinations by experts, revealed writing engraved on the wheels that referred to months of the year and the four-yearly Olympic Games. As the device was fragile much of it could not be dug out of the cement, so it was X-rayed. Recently a state-of-the-art X-ray machine took thousands of 3D images while it rotated on a turntable and this gave excellent pictures of small gear wheels that been hidden behind larger ones. The pictures allowed the number of teeth on the rim of the wheels to be counted. A model of the mechanism showed that by turning a handle the intermeshing gear wheels rotated at different speeds depending on their number of teeth. Each wheel tracked a different celestial cycle. For example, the wheel with 235 teeth displayed a nineteen-year calendar tracking the motions of the sun and moon. The teeth represented the 235 lunar months that equated to nineteen solar years. The cycle could be re-set repeatedly. Another gear had 223 teeth corresponding to the 223 months of the eighteen-year cycle of eclipses. There were twenty-seven surviving gear wheels, but originally there may have been twice as many. The instrument could predict the time of almost any celestial event. No device from antiquity remotely approaches the complexity and sophistication of the Antikythera mechanism. The inventor is unknown.

The Antikythera mechanism worked on the assumption of a centrally placed earth. Ptolemy's view went unchallenged until the sixteenth century. Nicolaus Copernicus, a brilliant Polish mathematician, was convinced that all the evidence pointed to the sun being at the centre of the universe and the earth was just one of the orbiting planets. He wrote his conclusions in a book entitled *Concerning the Revolutions of Celestial Spheres*. In it he stated that the apparent motion of the firmament was an illusion created by the rotation of the earth. He knew that his theory was blasphemous, so he kept the manuscript hidden. He held a copy of the printed book for the first time on his deathbed in 1543.

The Protestant leader Martin Luther was incensed by Copernicus's book and reminded readers: 'Jehovah ordered the sun, not the Earth, to stand still.' The Vatican banned the book and it remained so for 280 years. Ironically, Ptolemy had faked his data to fit his theory of a centrally placed earth.

Telescopes did not exist yet astronomers made major discoveries. A Danish astronomer, Tycho Brahe, was the greatest observational genius of his time. He measured the size and distance of celestial bodies and in 1572 was the first to observe an exploding star, what we now call a supernova. This didn't fit the prevailing belief that God had fixed the stars at the creation and they were immutable. He also found time for a duel over who was the better mathematician. *He* was, but his opponent was a better swordsman and sliced off his nose. Brahe replaced it with a gold nose glued on to the stump. He died in 1601, but had he lived just seven more years he would have been able to buy a device he would have swapped his golden nose for.

When spectacles were invented to rectify failing sight, lens makers proliferated. It was only a matter of time before someone put two lenses in a line and noticed that distant objects came closer. At the beginning of the seventeenth century there were several makers of spectacles in the Dutch town of Middelburg. More than one of them claimed to have invented the 'perspective glass', a tube with a lens at both ends and the ability to focus. But they thought it was a novelty toy with no practical use. The exception was a skilled German-Dutch lens grinder named Hans Lippershey who saw the military potential for spotting the enemy or reading a signal from a distant ship. In 1608 he offered his telescope to the government but they demurred, although they bought his other invention, binoculars.

A rebel with a cause

Galileo Galilei was the first to use a telescope to scan the night sky. He said: 'As I stinted neither pains nor pence . . . I obtained an excellent instrument which enabled me to see objects a thousand times as large.' He could see craters and mountains on the moon and although he perceived something jutting out from Saturn he couldn't distinguish it as rings. He nervously revealed that Saturn had ears. The larger moons of Jupiter were clear and circling the planet just as the earth's moon did. He told everyone he met about his discovery.

It was a mistake. Galileo was summoned to Rome to face the cardinals. They reminded him that *all* celestial bodies circled around the *Earth* and it was impossible for some to rotate around another planet. He invited the clergy to peep through his telescope but they 'steadfastly refused to look'. They warned him that: 'To affirm that the sun . . . is at the centre of the universe . . . is a very dangerous attitude.' His views were 'foolish, absurd and heretical'. He was dismissed on the understanding that he would only discuss the views of Copernicus as a hypothesis.

Galileo was an imprudent man. He wrote an article that concerned a debate between three men. One was a supporter of Copernicus, the second a supposedly impartial mediator but in fact an advocate of the Copernican theory and the third a well-meaning dullard called Simplico. The first man was persuasive: 'The Bible tells us how to go to heaven, not how the heavens go.' The Pope suspected he was being lampooned in the person of Simplico and Galileo was summoned to face the Inquisition. They found him 'vehemently suspected of heresy'. This was serious. Only thirty years earlier a man facing exactly the same charge had been burnt at the stake. Galileo was threatened with torture unless he recanted. As he finished his apologia, he couldn't resist whispering: 'Nevertheless it does move.' His sentence of life imprisonment was repealed by the Pope, but he spent the remainder of his life under house arrest.

He was dismayed that not just clerics but also professors damned him: 'As if I had placed those things in the sky with my own hands in order to upset nature.' However, on one point Galileo was proved wrong by the Vatican's librarian. Saturn did not have ears, its rings were in fact Our Lord's foreskin. Galileo was subsequently exonerated by Pope Paul II – 364 years too late.

The conventional 'Dutch' telescope was not ideal for astronomy. It had a narrow field of view and light passing through the lens was deflected and gave a hazy coloured halo around bright objects. Either different lenses or an entirely different type of telescope was required. As early as the thirteenth century an English monk, Roger Bacon, had predicted that, with the right combination of lenses and a *concave mirror*, 'the sun, moon and stars may be made to descend hither in appearance'. In 1668 Isaac Newton had the same idea. He made a working model of his reflecting telescope in which the light was focused, not by a lens but a concave mirror that reflected the light back up the tube to an eyepiece in the side. His model was only 6 in (15 cm) long but had a magnifying power of times forty, and there was no colour aberration. All subsequent astronomical viewers would be reflecting telescopes.

William Herschel, who was born in Germany and settled in London, was a successful musician and composer, and the greatest astronomer of the eighteenth century. He knew that to explore deep space and see the fainter, more distant astral bodies he needed a reflecting telescope with a wide tube to collect as much light as possible. As no lens grinders could supply him with bigger mirrors, he cast, ground and polished large, concave metal mirrors himself. His entire house was converted into workshops. After much experimentation he found that pounded horse dung provided the ideal non-porous mould for casting the mirrors. It was so successful that dung was still being used to cast giant mirrors for telescopes in 1920. On one occasion the casting exploded and white-hot metal poured across the basement floor and Herschel's workers leaped to safety. The most critical procedure was the final polishing of the mirror's surface. If it paused for even a few seconds the metal would mist over and the mirror was ruined. Herschel's sister Caroline wrote: 'finishing a seven-foot [2.1 m] mirror he had not left his hands from it for sixteen hours.'

He produced telescopes that were vastly superior to any other. Their power, clarity and brightness of images were unsurpassed. On his first night of scanning the heavens he discovered that the pole star was in fact two stars and in 1781 he discovered the planet Uranus, named after the Greek god who was the personification of the sky. Herschel went on to construct the biggest telescope in the world. Other astronomers couldn't believe that his telescope had a magnification of 6,450 times. His production of mirrors was prodigious. Over a single decade he made 350 mirrors for his 7-ft (2.1 m) and 10-ft (3 m) long telescopes and eighty for his 25-ft (7.6 m) device.

All this time Herschel was also composing and conducting orchestras all over England. In 1782 he became Personal Astronomer to King George III, whose only reservation was that Herschel was 'sacrificing his valuable time on crotchets and quavers'.

In those days large telescopes were erected outdoors supported by enormous trestles. In winter, viewing the sky was a bitter experience as mirrors iced up and the observer froze. Herschel rubbed his hands and face with raw onion to prevent frostbite, but surely this didn't leave his eyes in a good state for observing. One windy night the huge trestle collapsed, leaving Herschel buried beneath a pile of battens. A later Astronomer Royal, Nevil Maskelyne, invented an insulated suit for sky gazing. The breeches consisted of flannel lining, under a thick layer of wadding covered with brightly coloured silk. They covered the feet like a child's romper suit and there was a huge padded rear for sitting on a cold bench. His silken jacket was also quilted, like an eiderdown. The outfit may have kept Maskelyne warm, but few astronomers followed his fashion, for who would wish to look like an obese Michelin man with a prodigious rear end?

An invisible world

Despite our best efforts we haven't found any sign of life on a celestial body. If we had looked down instead of up we would have discovered hidden worlds teeming with life.

As with the telescope the window to this world was the lens. In 1853 a small oval of rock crystal, convex on one side, turned up in a collection of ancient Assyrian artefacts. Similar crystals have been found, sometimes in large numbers in archaeological digs. Some may have been used as burning glasses to light fires, but others seem better suited to use as a magnifying glass. One found in Crete thought to be 2,500 years old could magnify seven times.

A magnifying glass is now convex on both sides after being ground thinner on the edges than the centre. Galileo coined the name 'microscope' for a close-viewing device with two lenses, one at either end of a tube. The one nearest to the object being viewed enlarges the image and a second lens in the eyepiece enlarges it further. The arrangement effectively brings the viewer's eye very close to the object without any loss of focus. This 'compound microscope' was invented by another Middelburg lens grinder, Zacharias Janssen, around 1590.

The first scientist to exploit this 'novelty' for research was the Curator of Experiments at the Royal Society of London, Robert Hooke. He was brilliant but fiercely competitive and was forever feuding with fellow scientists over who had the priority on a new discovery. Although he had plenty of original ideas, he was not averse to claiming the ideas of others as his own. His flair for inventing was outstanding: often he solved a problem by making a new invention.

Under the microscope Hooke examined a variety of small creatures, including fleas and lice, which were well known to everyone, although they hadn't seen them looking quite so frighteningly large. In 1665 Hooke published *Micrographia*, a compendium of his finds with magnificent fold-out pages of engravings. He had trained as an artist, so his original drawings were immaculate. He was also a meticulous observer: 'I never began to make a draft before, by many examinations in several lights in several positions . . . I had discovered the true form.'

Samuel Pepys bought Hooke's book and marvelled at the strange, compelling creatures into the small hours. The coffee-table book of its time, *Micrographia* was a best-seller and the microscope became a must-have toy for wealthy gentlemen.

One day Hooke cut a thin slice of cork and placed it under the microscope. He saw 'a great many little boxes', which he called 'cells' as they were reminiscent of a monk's cell. He was the first person to see the cellular building blocks of living things in which DNA is stored and all the chemical reactions of life occur.

Antonie van Leeuwenhoek, a Dutch draper with no scientific training, was thrilled by Hooke's book and took up microscopy. He made his own device, which was a paddle-shaped board with a hole in which he embedded a home-made lens. He didn't grind his lenses; instead he warmed glass, drew it out into a fine thread and then melted one end to produce a tiny transparent bead. That was his lens. Hooke had a state-of-the-art compound microscope with the finest ground lenses that magnified thirty times. Leeuwenhoek's glorified magnifying glass was nine times more powerful.

At the time the prevalent view was that new life was created by spontaneous generation: maggots were spawned from rotting meat, weevils from stored grain and eels from the dew. Leeuwenhoek observed that insects laid eggs that hatched into baby insects or maggots. With his high magnification he could see organisms that were invisible to the naked eye. Wherever he looked, in rainwater, urine, even his saliva, he found millions of frantically dancing 'animacules'. He also examined his blood cells and semen, which, he hastened to add, 'was the residue of conjugal coitus'. He suggested that the spermatozoa penetrated the egg to consummate fertilisation. Later, microscopists less observant but more imaginative than Leeuwenhoek, reported that every spermatozoid contained a 'homunculus' – a tiny man. Females therefore had no role in reproduction except as an incubator.

Leeuwenhoek couldn't draw but he was a close friend of the painter Vermeer, and indeed executor of his estate. Probably through Vermeer's good offices he was able to get artists to look down his microscope and draw what they saw, provided it was the same thing that he could see.

At regular intervals Leeuwenhoek sent communications to the Royal Society in London for publication. They were not scientific articles, but a list of his latest observations. The first submission was accompanied with a humble apology: 'Take my simple pen, my boldness and my opinions for what they are, they follow without any particular order.' To make matters worse they were written in Dutch and had to be translated by the Society's secretary and Hooke, who learned Dutch to do so.

Leeuwenhoek was a great amateur scientist who opened our eyes to a secret and unsuspected world. His tiniest find in 1683 was his greatest discovery: clumps of unprepossessing microbes that would later be called bacteria. He suggested that they might be dispersed by the wind. Daniel Defoe, in his *A Journal of the Plague Year*, enlarged on the wind's influence: 'by carrying with it vast numbers of . . . invisible creatures who enter the body with the breath or even at the pores . . . and there generate or emit most acute poisons . . . which mingle themselves with the blood and so infect the body'. It took almost two hundred years to prove that he was right.

THE GERM OF AN IDEA

'Diagnosis: a physician's art of determining the condition
of the patient's purse, in order to find out how sick he is'
– Ambrose Bierce

L ike Defoe, medical men of his day believed that the wind
had a hand in transporting disease. They also believed
that typhoid fever came from bad smells and malaria (which
means 'bad air') was a miasma rising from damp soil. When
cholera blew in, whole towns were quarantined and the
authorities filled the air with smoke from tar barrels and fired
cannons into the sky to disrupt the miasma. 'Experts' claimed
that cholera was a miasmatic electric effluvium because, like
lightning, it struck suddenly, blackening its victims. The
answer was to galvanise the patient. One poor woman was
shocked for seven hours.

A popular saying was: 'Ask not of what disease he died,
but of what *doctor* he died.' For all the fancy vocabulary of
the expensive physician and the cure-alls of the boastful
quack, no one knew what diseases really were or how to cure
them. And only the quacks knew they were charlatans.
Immanuel Kant wrote: 'Physicians think they do a lot for the
patient when they give his disease a name.'

The speckled monster, smallpox

Prevention, it is said, is better than cure and that is certainly true of smallpox, which the English historian Macaulay called 'the most terrible of all the ministers of death'. In cities it carried off one in three children. Even royalty were not immune. Among its victims were Louis I of Spain, Peter the Great, Louis XV of France and the Duke of Gloucester, who died aged eleven, ending the Stuart dynasty.

The ancient Chinese blew flakes from smallpox scabs up the noses of the healthy in the hope that it might protect them from the disease. As it sometimes gave the patient smallpox, the scabs were dried out to weaken their potency. A different means to the same end existed in Turkey in the eighteenth century.

Lady Mary Wortley Montagu, a wise and witty London socialite, lost her brother to smallpox in 1713. Two years later she also fell foul of the disease, which ravaged her beauty. Subsequently she spent two years in Constantinople while her husband was the British Ambassador. She wrote home: 'The smallpox, so fatal and general among us, is here entirely harmless by the invention of engrafting.' This involved scratching the skin with a needle and then smearing on it matter from the pustule of a person with a mild case of smallpox. Lady Montagu was so impressed that she had her five-year-old son treated by a Greek matron armed with a rusty needle who brought the 'fresh matter' in a nutshell. Her ladyship didn't inform her husband until the boy's fever had subsided.

On her return to England she extolled the benefits of inoculation and her daughter was the first person in Britain to be immunised in this way. She persuaded Princess Caroline, daughter of George II, to have her children treated, but she agreed only after eleven orphans were inoculated to show that it was safe.

The technique was to transfer pus from someone with smallpox and using a live germ this way was not entirely risk-free. Indeed George III's son Octavius died after an inoculation. Nevertheless, a new trade of inoculating developed and the most famous proponent was Thomas Dimsdale. In 1768 he was invited to introduce inoculation into Russia. When he arrived he was shocked to learn that his first patients would be Catherine the Great and her son the Grand Duke. He was petrified of what might happen to him should anything go wrong. He was informed that the mother of the boy who was to provide the pus was hysterical because she believed that a donor would certainly die. It was rumoured that Dimsdale had a coach and horses standing by should he need to make a quick getaway.

But Catherine's treatment went well and later her son was inoculated with matter from her sores to convince others that donating was not deadly. Dimsdale spent several weeks inoculating grateful aristocracy. One gave him such a heavy pouch full of gold roubles that he could hardly shuffle out of the room. Catherine made him a baron and gave him £10,000 plus an annuity of £500 for life. He returned to England a very wealthy man. So he ditched medicine and bought a bank.

In 1770 Edward Jenner, a medical student at St George's Hospital in London, entered the crowded smallpox ward, where pitiful patients were packed four to a bed. It was a sight he would never forget. As usual there was a smallpox epidemic. The following year the French army fighting the Prussians lost twenty thousand men to smallpox, vastly more than were killed in action.

After graduating Jenner established a country practice and was also a consultant in Cheltenham. He spent his spare time bird watching and would later be elected to the Royal Society for unravelling the murderous behaviour of cuckoo chicks. He also studied milkmaids, as men often did, and noticed that none of them was scarred by smallpox, although they caught cowpox, a much milder disease with similar symptoms. He wondered whether the cowpox had enabled them to fend off smallpox. The advice of his teacher, the distinguished Scottish surgeon John Hunter, was: 'But why think? Why not try an experiment?' Jenner heard of a local farmer who thought along the same lines and used a darning needle to give his family cowpox.

Jenner persuaded Sarah, a dairymaid who had caught cowpox from a cow called Blossom, to allow him to take pus from one of her sores. This he scratched into the arm of a healthy eight-year-old boy, who developed a light fever. Six weeks later he injected the lad with smallpox, saying: 'Don't worry, little boy, it's just a scratch.' Whether his parents had given permission for this bold and dangerous experiment is unknown. Luckily, the boy recovered. He seemed to be immune, but to make sure, Jenner gave him a second shot of smallpox. It is good to report that the boy lived well into old age.

Jenner wrote an article describing his experiment, but the Royal Society rejected it. So he wrote a controversial book detailing twenty-three successful vaccinations. It sold well and his cowpox matter was widely available. Having read

Jenner's book, Louis Pasteur coined the word 'vaccination' (from *vacca*, Latin for 'cow').

There was widespread resistance to the very idea of vaccination. Scottish Calvinists claimed it challenged Divine Providence. One detractor called it 'a loathsome virus from the blood of a diseased brute'. To test its efficacy felons in London's Newgate Prison became guinea pigs, with a pardon for the survivors. In the United States the Boston Board of Health held a public demonstration. Nineteen volunteers were given cowpox and then smallpox. Two 'control subjects' received just smallpox. The trial was 'completely successful'; in other words, only the controls died.

Jenner had famous supporters: the writer Fanny Burney, who had her baby vaccinated, and her friend the artist Thomas Gainsborough. President Thomas Jefferson had his family and his neighbours protected. As a baby Queen Victoria was also treated.

In Britain there was no mass vaccination programme for the population until it was made compulsory in 1853. Failure to comply could incur fines, seizure of goods, even imprisonment. These draconian measures spawned a large lobby against vaccination. Protesters railed against 'medical spies forcing their way into the family circle'. The term 'conscientious objector' was first used in 1898 of parents who refused to have their children vaccinated. A recruitment poster for the National Anti-Vaccination League pointed out 'the unparalleled absurdity of deliberately infecting a healthy person . . . with the POISONOUS matter of a DISEASED CALF'. Some intellectuals rallied to the cause but their contributions were neither informed nor enlightening: 'a particularly filthy piece of witchcraft' (George Bernard Shaw) and 'vaccination, a giant delusion . . . never saved a single life' (Alfred Russel Wallace, the co-proposer of the theory of evolution). Malthus did not approve of vaccination because smallpox was God's method of population control.

Vaccination saved millions of lives from a disfiguring and often fatal disease. A hundred years on, smallpox was the first, and as yet the only, disease to be eradicated. Except for a few strains kept in a couple of laboratories. What for, one wonders?

Doctors of death

Despite the urgency to find the causes of disease, sometimes the medical establishment failed to accept the obvious cause. Of the numerous diseases endemic in hospitals perhaps the most heart-rending was childbed fever, which killed mothers within days of having delivered their baby. In 1795 Alexander Gordon, a Scottish physician, and forty-eight years later Oliver Wendell Holmes in the United States both came to the conclusion that childbed fever was transmitted to the mother by the attending staff. But nobody took any notice.

Ignaz Semmelweis was a Hungarian physician working in Vienna. In his hospital there were two maternity wards and in 1846 he noticed that the death rate for childbed fever was almost ten times greater in the ward attended by physicians compared to that staffed by midwives. He also noted that the students came directly from the post-mortem theatre to the delivery bed without washing their hands. Could they inadvertently bring something with them? Something deadly?

Semmelweis instigated a strict regime of washing hands and surgical instruments in bleach. The mortality rate in the physicians' ward plummeted to only 1.27 per cent. Doctors were appalled, as they thought of themselves as God-like healers, not angels of death. Semmelweis's contract was not renewed, so he left medicine to help Schliemann to excavate Troy. An eminent obstetrician listed thirty possible causes of childbed fever that did not include Semmelweis's explanation. It would be almost two decades before there was a major innovation in surgical practice.

The germ killer

Joseph Lister spent an idyllic childhood in Upton Park, now the home of West Ham United's football ground. His father was a wine merchant who loved microscopy and was a self-taught expert in optics. So expert in fact that he solved the vexing problem of colour aberration in microscope lenses and was elected to the Royal Society.

When he was fourteen Joseph spent his summer holiday dissecting a sheep and labelling the cranial bones in a human skull. Occasionally he and his brother would sneak down to the banks of the Thames, where pirates and cattle thieves were left dangling on a gibbet until they decomposed. Although he was too late to save the hanged felons, Lister would spend his adult life trying to prevent putrefaction. As a medical student he honed a pathological inability to be punctual. But surgery got him there on time: 'I am from day to day experiencing in this bloody and butcherly department of the healing art . . . I almost question whether it is possible such a delightful pursuit can continue.' What, I wonder, was the patient's view of this 'delightful' experience?

In 1858 Lister secured a junior post at Edinburgh University

Medical School. His good looks seduced the chief surgeon's daughter, Agnes. They spent their honeymoon touring the medical research centres of the continent and discussing his favourite topics – inflammation and putrescence. It was something of a pusman's holiday. If nothing else, it prepared Agnes for what to expect. Back home Lister experimented on rabbits and a dead cow, leaving little room in the kitchen for food preparation. He also drained blood from a horse to mix with – you've guessed it – pus. Agnes wrote down his methods and observations in a recipe book. He also invented surgical aids such as the silver wire suture needle and specially treated slow-dissolving catgut so that stitches could be left inside the patient.

Lister was elected to the Royal Society for his work on inflammation. One of his findings was that if you removed the brain from a frog its arteries didn't get engorged. Unfortunately, brainectomy was not a practical cure for inflammation.

With his father-in-law's support he was appointed Professor of Surgery at Glasgow Infirmary, which was built on the burial pits of cholera victims. Like many surgeons of the day, he was dismayed when patients survived surgery only to succumb to St Anthony's fire (gangrene) and other forms of tissue decomposition. If it was disheartening for the surgeon, it was also a bit disappointing for the patient.

Lister wrote of 'the disastrous consequences in compound fracture contrasted with the complete immunity to danger to life and limb in simple fracture is one of the most . . . melancholy facts in surgical practice'. Said simply, if the skin is unbroken, bone fracture is a minor problem, but if the bone breaks out through the skin, putrefaction almost certainly follows. The surgeon's response to a compound fracture was therefore amputation of the arm or leg. The average number of amputations at Glasgow Infirmary was forty-three per cent and the death rate of twenty-five per cent was regarded as 'satisfactory'. What was going wrong?

Sir Frederick Treves, the surgeon who rescued the 'Elephant Man', Joseph Merrick, recalled the state of the ward when he was a student: 'Cleanliness was out of place, it was considered to be finicking and affected. An executioner might as well manicure his nails before chopping off a head . . . Suppurating wounds were washed daily but only one sponge on each ward. With a daily wash with this putrid article . . . any chance that the patient had of recovery was eliminated.'

And what was the causal agent? Frenchman Louis Pasteur was a budding artist who became a chemist. He didn't shine at school but his mind could make imaginative leaps. He had shown that heated sterilised broth did not go rancid in a flask sealed only with a cotton-wool bung. Removing the bung led to the broth turning putrid, so something in the air triggered putrefaction. He suggested that tiny microbes were floating in the air.

Lister was sure that these microbes were landing in open wounds and causing them to suppurate. He had one of his father's latest microscopes, the only one in the infirmary. He took a swab from a wound and saw hosts of living microbes. So he searched for something to kill them.

In the past wine and vinegar were the only disinfectants – remember how, in the nursery rhyme, Jack's crown was treated with vinegar and brown paper? Lister needed something stronger and tried many caustic compounds, even nitric acid, the last of which would certainly kill germs while dissolving the patient. He read that in Carlisle the local authority used phenol (carbolic acid) to 'sweeten' sewage. It was produced from creosote and its main use was for preserving corpses. Across the English Channel physicians encouraged victims of cholera to drink a phenol cordial when what would have helped was a jug of unpolluted water.

Lister used full-strength phenol in large quantities to wash out wounds and clean the surrounding skin. He also invented a spirit lamp that sprayed phenol to shoot down airborne germs; the students called it 'let us spray'. After two disappointing trials, especially for the patients, he operated on a boy whose leg had been run over by a wagon wheel. Although the bone had broken through the skin, the wound was very small, which would give the best chance of success. It went well: six weeks later the boy was able to hobble from the hospital, although the phenol had seriously burnt his skin. Lister boasted that his phenol treatments caused the mortality rate to fall from forty-six per cent to fifteen per cent, but although he had always dismissed the benefits of hygiene, it probably contributed significantly to his improved figures. In his last three years at Glasgow not a single patient died from post-operative infections.

Lister's work was damned by other surgeons as being based on an unproven theory. They didn't fancy the effort of cleaning wounds and applying ever more complicated dressings and trying to operate in a caustic cloud of phenol. Not surprisingly, they preferred the old way of just striding into the theatre without washing their hands and sawing off a limb or two.

But resistance slowly declined. The future was not even antiseptic: it would be aseptic. The operating room would be germ free, with sterilised instruments, spotless surgeons and nurses wearing rubber gloves. Lister was heralded as 'creating anew the ancient art of healing'. He is also commemorated in Listerine, the antiseptic mouthwash. A probably apocryphal story has it that the company came up with the product and the name and then got an advertising agency to devise a condition for it to cure.

Defeating disease

Meanwhile Pasteur had concentrated on commercial processes: the souring of wine, the fermentation of alcohol and the origin of carbon dioxide, which made bread rise. He discovered that these were not simply chemical reactions. His microscope revealed a myriad of tiny micro-organisms that caused fermentation and he discovered they could be killed by high temperatures, in the process we now call pasteurisation.

By 1878 Pasteur was convinced that fermentation, putrefaction and disease were all caused by microbes. That same year Robert Koch, a German country doctor, wrote an article outlining germ theory. He listed six diseases and named the six bacteria responsible for them. He also laid down the rules for positively linking a specific bacterium to a disease. The germ theory was no longer a theory.

It was an astonishing breakthrough and it came about because Koch's wife bought him a microscope for his birthday. No gift in history has been so significant. With it he was able to differentiate bacteria by their size and shape. He attached a camera to his microscope and photographed them as if they were mug shots of suspected criminals.

When the local sheep were dropping dead in the field Koch

isolated the anthrax bacterium from them. One day he strode into his surgery with a large jar full of eyeballs because he thought that the sterile fluid inside the eye would be ideal for culturing bacteria. From his cultures he discovered that the anthrax bacterium could go dormant and persist in the soil to trigger an epidemic later on.

Koch's insight was recognised and he was given a laboratory and the funding to run it. The first task was to perfect methods for culturing bacteria. The ideal receptacles were shallow glass Petri dishes with a lid, named after Koch's assistant Julius Petri. These were half-filled with gelatin to promote the growth of microbes. Within a year Koch was able to write: 'I consider it as proven that in all the tuberculosis conditions . . . there exists a characteristic bacterium which I have designated as the tubercle bacillus which . . . can be distinguished from all other micro-organisms.'

Meanwhile Pasteur was attempting to make vaccines against diseases. He studied chicken cholera and discovered that if the serum from a chicken with mild symptoms was given to a healthy chicken, and this process was repeated from one chicken to the next, fifteen times, the cholera was sufficiently weakened to use in a vaccine that 'preserved them from the deadly form'.

Using a similar procedure Pasteur was able to produce a vaccine against anthrax. The Agricultural Society challenged him to demonstrate his vaccine and provided sheep for the trial. This was just the publicity he was looking for. Twenty-five sheep were vaccinated and then infected with anthrax, while another twenty-five were infected without prior vaccination. A month later all the vaccinated sheep were fine and all the unvaccinated ones were dead. The experiment made Pasteur's name, but it transpired that for some reason he hadn't used his vaccine but one developed by Jean-Joseph Toussaint.

Three years later Pasteur developed a vaccine for the dreaded rabies. A ten-year survey revealed that of six thousand people bitten by rabid dogs less than one per cent of those who had been vaccinated died.

Back in Germany Koch was 'never idle'. When he wasn't mountaineering, digging up ancient artefacts or playing the zither, he was identifying another dangerous microbe. In 1884 he found the comma-shaped cholera bacterium and in the following years his team tracked down the causal agents responsible for bubonic plague, leprosy, meningitis, pneumonia, syphilis, tetanus, typhoid and whooping cough. His most dramatic moment concerned a boy with diphtheria, a disease in which a leathery membrane of dead cells grows over the throat, blocking the airways until the victim suffocates or is killed by its deadly toxin. Koch rushed to the boy's bedside on Christmas Day 1895 to give him the antitoxin he had developed. It saved the boy's life and Koch became famous overnight.

He found that for some diseases, such as tuberculosis, he was not able to develop a vaccine or antitoxin. Others less able thought they had the answer. Englishman John Pugh heard of a TB sufferer being taken up in a balloon and came down feeling better. So he proposed using balloons to carry heavenward 6,000 ft (1,830 m) joined-up lengths of metal tubing down which 'pure life-giving air free from microbes' would flow. The lower end of the pipe would be jammed in the 'lofty ceiling' of a building occupied by a couple of thousand patients at a time. To Pugh's dismay the building was never built.

Many conservative doctors refused to relinquish their belief in miasmas. One declared: 'I care not a fig for germs,' while another went so far as to drink a culture of the cholera virus to prove it was harmless. His extraordinary story is related in my book *Smoking Ears and Screaming Teeth*.

Serendipity

Both Koch and Pasteur were honoured all around the world. They had invented bacteriology and revolutionised medicine. In following years every major hospital came to have a bacterial research laboratory. At St Mary's Hospital in London some of the work focused on the *Staphylococcus* bacterium, which was responsible for pneumonia, abscesses and blood poisoning, one of the post-operative infections that Lister had battled with. One of the staff was Scottish bacteriologist Alexander Fleming. He had been invited to work at St Mary's because he was a crack shot with a rifle and the other researchers were determined to win a trophy at Bisley. He had been a brilliant student but he was a poor teacher and a dull conversationalist.

In 1921 Fleming had a heavy cold and rather than waste the mucus running from his nose he plated it on to a Petri dish. His colleagues all called him 'Flem', although it may have been 'Phlegm'. He found that the mucus killed some of the bacteria. He attributed this to an enzyme he called lysozyme. As he was no chemist he failed to isolate the enzyme, so instead of getting help he gave up. According to a co-worker, the agar plate on which he had smeared the mucus was spotted with bacterial patches except where the mucus was. The bacteria had clearly not been deliberately smeared on the plate: they had arisen from several bacterial spores settling on the agar when the lid was off. If so, that was exceedingly careless.

In 1928 almost the same thing happened again, but this time it would make Fleming famous. His version of events was that he had plated up some *Staphylococcus* swabs from a patient and incubated them over the weekend. On the following Monday he examined them and found one was contaminated with a large patch of bread mould. He thought it odd that the bacterial colonies had not grown in proximity to the mould.

The contaminating mould was identified as a species of *Penicillium* (Latin for 'paintbrush'). As before, Fleming had no idea how the mould had got on to the plate. It had probably originated from the lab below which had been culturing this very strain. It could have drifted out through the door, which was always kept open, and into Fleming's lab, which also had an open-door policy, finally landing on an uncovered plate. If so, this smacks of shockingly lackadaisical practice.

Fleming gave the name 'penicillin' to the anti-bacterial secretion that came from the mould. Although he cultured the mould and tried to extract the penicillin he had no success. He had researched alone for his entire working life, so instead of inviting experts to extract and isolate penicillin, yet again he abandoned the project.

He wrote a brief article on his discovery in which he characterised penicillin as a lytic (cell-eating) enzyme. It was neither of these things, but these errors led to other scientists becoming interested in penicillin. Howard Florey was an Australian pathologist and Ernst Chain a brilliant biochemist who had escaped from Nazi Germany. They met at Oxford

University and joined forces to find a disease-killing bacterium that didn't damage the patient's cells. They began by searching the scientific literature for lytic enzymes and came across Fleming's article.

They scratched around for funding but the British Medical Research Council wasn't interested. However, the Rockefeller Foundation stepped in. Florey and Chain soon extracted and purified penicillin with help from Norman Heatley, a British chemist who was a whiz at stabilising and bulking up the yield of penicillin. By 1940 they had enough antibiotic to begin trials. They injected eight mice with *Staphylococcus* and treated four with penicillin. Only the latter survived. Had they used guinea pigs rather than mice the trial would have failed – guinea pigs are allergic to penicillin – and the research might had stopped right there.

It was time to risk a test on a human. The patient was a policeman who had been scratched by a rose and had already lost an eye from the spreading infection. With penicillin he was recovering until the supply ran out and he relapsed and died. The problem was that a man is three thousand times heavier than a mouse and therefore needs vastly more penicillin. With more supplies they treated a young lad suffering with septicaemia. He recovered, only to succumb to a ruptured artery unconnected to the penicillin treatment.

With half the world at war it was imperative to get this life-saver into mass production. All the British pharmaceutical companies were overworked, so Florey and Heatley flew to the United States. Florey's overcoat had been soaked in Penicillium from which new cultures could be started. Trials on wounded patients proved that the antibiotic could render potentially fatal infections harmless. To increase productivity they searched for a strain of *Penicillium* that produced greater amounts of penicillin. The found the mould they sought growing on a melon in the local market. By D-Day there was sufficient penicillin to treat all the Allied troops.

As a favour Florey gave some penicillin to Fleming to treat a friend and it saved his life. The press acclaimed Fleming, who apparently had 'worked tirelessly' to bring this project to completion. Fleming cultivated the press more energetically than he had the mould and made no mention of the others who had transformed his chance observation into the most powerful drug the world had ever seen. It halted the growth of bacteria that caused blood poisoning, gangrene, meningitis, pneumonia, syphilis, gonorrhoea, scarlet fever and dozens of less serious diseases. It also stimulated the search for other antibiotics.

In 1945 Fleming, Florey and Chain shared the Nobel Prize for Medicine and all three were knighted. Norman Heatley, whose skills were the key to their success, was overlooked. He had to wait until 1990, when Oxford University recognised his contribution and gave him an honorary degree.

A DIRTY BUSINESS

'What happens to sewage effluent after it has been treated?'
– exam question

'It's used to make toilet water'
– student's answer

Although surgeons were reluctantly beginning to wash their hands, their patients were probably not washing at all. Bathing had a chequered history in Europe. It was sometimes fashionable with the well-to-do, or water might be shunned.

Bath time

The royal palaces in ancient Mesopotamia, the Indus Valley civilisation and the Minoans in Crete all had bathtubs. The Romans went in for communal bathing and some of the grandest buildings in Rome were public baths. Large aqueducts brought 38 million gallons (173 million litres) of water into the city every day. Vast amounts of water were needed to supply the numerous fountains and the public baths, where a pipe led to the boiler to be heated and then pumped through the world's first underfloor central heating system. In some

271

rooms the floors were too hot to cross with bare feet. Smoke from the boiler fire warmed the hollow walls.

Curiously, the water baths were not for washing. Instead the men rubbed themselves with sand and oil and scraped these off with the underlying dirt. A celebrated gladiator sold samples of his scrapings to female fans who used it for face cream. The communal bath was chiefly a place to make deals, crack jokes and swap gossip, and was also renowned for drunkenness and sexual excess. In some cases there was no hole to allow water to run away and the fermenting 'soup' was just topped up, not replaced. Marcus Aurelius said: 'What is bathing when you think of it – oil, sweat, filth, greasy water, everything revolting.'

Christians were not encouraged to wash. St Jerome had said: 'He who has bathed in Christ has no need for a second bath.' In consequence, St Thomas Aquinas encouraged the use of incense burners in church to mask the aroma emanating from the congregation.

The Middle Ages saw a resurgence in communal bathing all over Europe. The notion of getting washed was soon eclipsed by the thrill of getting naked, except for a hat, of course, and sharing a cosy two-seater tub with someone of the opposite sex. Even monks, priests and nuns threw off their habits and got into the spirit of things by tubbing with a nude companion eager to discuss the scriptures.

Some women insisted on having a clause in the marriage agreement that required her to go to the baths twice a week for her 'health'. An Italian proverb advised: 'If you want your wife to conceive, send her to the baths and stay home yourself.' Even the trollops complained that bored wives were ruining their trade.

The prevalence of plagues did for the baths, which were viewed as cauldrons of contagion. The aristocracy had private baths but they were little used. Queen Elizabeth I was considered to be a fanatical washer. She took a bath once a month 'whether she needed it or no'. In the seventeenth century anyone who bathed was considered eccentric. When the Marquis de Condorcet revealed to an aristocratic lady that he bathed occasionally, he was asked: 'Were you born under the sign of fishes?' Louis XV of France was said to take only one bath a year. Cosmetics and perfumes replaced washing.

Eighteenth-century spa towns revived the idea of bathing for your health. The writer Tobias Smollett was unimpressed with the level of hygiene at Scarborough's baths, which contained: 'Sweat, dirt and dandruff and the abdominal discharges of various kinds.' There were more exclusive baths in which men bathed upright accompanied by a floating bowl full of life's necessities such as snuff and face powder should one's make-up be compromised by a tiny wavelet.

The nineteenth century saw the building of public baths in every city. The Victoria public baths in Manchester had three pools. Fresh water was pumped into the First Class Men's pool every three weeks. Its old water had been pumped out, filtered and sent to the Second Class Men's pool. Its old dirty water had been filtered and pumped into the Women's pool. What could be fairer than that?

At the homes of the less well-off there was a similar regime. They dragged a tin bath in front of the fireplace. Father always bathed first, mother came second and the children followed in descending order of age. By the time it came to the youngest, the water was so opaque that throwing away the baby with the bath water was a distinct possibility. Tin baths were still commonplace in the 1930s.

Medics now encouraged cleanliness and were dismayed by the public's 'dread of water'. Perhaps they had been reading a manual of hygiene that warned: 'Certain parts must not be washed more than once a day or it might lead to terrible consequences.' But which parts?

The Great Midland Hotel in London epitomised the value of baths in the British psyche. Built in 1883, it was luxurious, with three hundred bedrooms but only five bathrooms. Even by 1950 some houses were still without a bathroom or hot running water.

For those not satisfied with an ordinary bathtub there were medicated baths with oxygenated water containing ozone, peroxide and malt extract. Better still there was a wave-producing tub. The bather sat on a trolley and pulled a spring back and forth, creating a 'powerful wave action', which would probably flood the bathroom floor and bring down the ceiling of the room below. The inventor added that the trolley could be replaced by a swing.

Please pull the chain

Washing was one thing, but washing away our waste products was a greater challenge. Four thousand years ago the royal palace at Knossos in Crete boasted a lavatory beneath which a stream acted as a flush. The ancient Romans, a nation of sanitary engineers, had communal toilets with twenty or so holes in a long stone slab on which the occupants sat shoulder to shoulder. In lieu of toilet paper there was a wet sponge on a stick, hence the expression 'to get hold of the wrong end of the stick'. The Romans brought sanitation to their empire but once they left, the locals forgot how to wash and returned to squatting in the fields. In medieval Britain ordure was called 'gong', not to be confused with a call for dinner.

The first manually operated flushing toilet was the invention of John Harrington, an amateur poet, professional ladies' man and godson to Queen Elizabeth I. He was banished from court for his scandalous gossip, so he spent his time building a house and perfecting his lavatory. His patent application was refused on the grounds of 'impropriety'. But the Queen tried it and ordered one.

Harrington wrote a book entitled *The Metamorphosis of Ajax*. Ajax was a pun on 'jakes', a slang word for a privy. He detailed the mechanism of the flush and boasted that with his mechanism 'unsaverie places may be made sweet . . . filthie places made cleanly' and its flush came 'with some pretie strength when you let it in'. The cistern above supplied the water and in his illustration it housed a live goldfish. A recent reconstruction of his device showed that it worked quite well. The only drawback was that it was designed to flush after twenty or so uses.

For other folk there was only the outside privy and the chamber pot in a cupboard, behind a curtain in the dining room or under the bed. For stability one bedpan had large, sharp teeth 'to engage the bedclothes' but it ripped them to shreds. In the morning the contents were thrown out of the upstairs window with a call of '*Gardez l'eau!*' to warn passers-by below. *L'eau* (water) probably gave us the word 'loo' for toilet. The householders were not the street cleaners, if indeed there were any.

Privies drained into cesspits that invariably overflowed. Samuel Pepys records in his diary: 'Going into my cellar . . . I put my foot on a great heap of finds . . . Mr Turner's house of office is full and comes into my cellar which doth trouble me.'

Even the capital of England didn't have a reliable source of clean drinking water until 1613 when Hugh Middleton made the New River, 40 miles (64 km) long and flowing through hollow wooden pipes southwards into the City of London. Lead pipes took the supply to individual houses, but each residence received water only on an allotted weekday.

In 1775 Alexander Cummings, a London watchmaker, patented a flush toilet with a 'stink trap' (a U-bend that prevents odour). Two years later Samuel Prosser invented the ball valve, which automatically refilled the cistern and cut off the water when it was full. The name 'water closet' was coined for the indoor lavatory and marked the change from outdoor privy to indoor privacy. The French called it '*un lieu à l'anglaise*' (an English place). However, the flush was embarrassingly noisy and some people used the chamber pot and then emptied it into the lavatory.

At the Great Exhibition of 1851 in London thousands of visitors were astonished at the demonstration of the flushing loo. Flush toilets became the must-have invention and they were fine if there was somewhere to flush to. Soon there were some 200,000 water closets in the capital and the same number of overflowing cesspits. Windsor Castle boasted fifty-three overflowing cesspits causing blockages in the sewer pipes that served it. These were cleared by 'toshers', young boys who crawled inside the fetid, gas-filled and rat-infested sewer.

John Snow, the doctor who kept Queen Victoria amused with chloroform during childbirth, was also a pioneer in public health. He traced the origin of a cholera outbreak to contaminated drinking water. But the authorities in London were convinced that cholera was wind-dispersed, so they ignored his findings and cholera returned to the city several times. Only engineers saw that sanitation made a difference because 'the places formerly most favourable to the spread of disease became quite free from it when afterwards properly drained'.

Sewage had rendered the River Thames devoid of all life except for bacteria. In the hot summer of 1858 'The Great Stink,' which even the most heavily perfumed handkerchief couldn't mask, closed Parliament. Something had to be done and the task was given to the Chief Engineer to the Board of Works, Joseph Bazalgette. He was a very small man with a large moustache, so he looked as if he were peering from behind a bush in the garden, but he was a dynamo of energy. It took £3 million and ten years to lay 1,200 miles (1,930 km) of ceramic sewer pipes and to build the 3.5-mile (5.6 km) Chelsea, Victoria and Albert embankments. They housed the massive collecting sewer that would carry away the effluent to be released downstream. These improvements added twenty years to the life expectancy of Londoners.

It would be some time before sewage treatment came in, or even screening to remove any foreign objects. According to a student in a recent examination, screening removed 'large pollutants like grit, dead sheep and canoeists'.

With adequate drainage the sale of flush toilets blossomed, in one sense quite literally, for they bore transfer or hand-painted pictures of exuberant chrysanthemums and interlacing vines. No decoration was too ostentatious and the lavatory pedestal became known as 'the throne'. The bathroom was also invented to accommodate the sanitary furniture.

And so we come to Thomas Crapper, the man who didn't invent the flush toilet and did not have crap named after him. 'Crap' is a very old word for chaff, the part of the cereal that is thrown away. Crapper was, however, a remarkable man. When only eleven years old he trudged the 185 miles (300 km) from Doncaster to London. Within fourteen years he was a master plumber and the owner of his own company. He tested every flush – if it could shift apples it could move anything.

Many of the lavatory cisterns dribbled all the time and new regulations insisted that lavatories had to conserve water. Alfred Giblin invented the Silent Valveless Water Waste Preventer. Crapper bought the rights and the lavatories sold like hot 'jakes'. 'Thomas Crapper' was printed on every appliance he sold, so he was truly a household name. His lavatories were also given impressive names such as 'The Demon', 'The Deluge' and 'The Dreadnought'.

Crapper plumbed Sandringham House when it was bought by the Prince of Wales. He also invented the manhole cover, new types of drains and pipe joints, but his great innovation was the bathroom showroom in Chelsea that boosted sales even though delicate ladies sometime swooned at the sight of 'unmentionables'.

Long after he retired, he was woken one night by someone having difficulty trying to make a flush work. He got up, fixed the problem and returned home to his deathbed. That's the kind of man he was.

Flushed with success

The first municipal public lavatory was opened in London in 1853. It cost only one penny and remained so for 118 years until the decimalisation of the currency. Hence the expression 'to spend a penny'. The father of Nevil Maskelyne, who you will remember embarrassed Marconi by breaking into his 'secure' transmissions, invented the penny-taking locks on pay toilets. Another Englishman, Mr A. Ashwell, patented the first practical 'Vacant/Engaged' bolt in 1882 and Thomas Crapper devised the automatic intermittent flush for urinals. Labelling the toilet doors raised even the most uncouth to the status of Ladies and Gentlemen. Nowadays we have been reduced to silhouettes of Little Bo Peep and a gingerbread man.

We now come to the delicate topic of bottom-wiping. Roman gladiators carried two sponges at all times. One was for drinking from a stream and the other for wiping his bottom. Let us hope they always remembered which one was which.

The British shunned the bidet because it came highly recommended by Parisian prostitutes, so we came to rely on paper. Paper was on bottoms long before it was in books. The Chinese manufactured it almost two thousand years ago and even the coarse and brutal Emperor Hongwu purchased fifteen thousand sheets of soft, perfumed toilet paper. In the Middle Ages British troops used sheep's wool or any available leaves except holly. The rich had an 'arse-wipe' attendant to do the dirty work for them. The bottom line was that his prospects were very low.

In the nineteenth century Joseph Gayetty, an American, was the first to manufacture medicated toilet paper, which he trumpeted as the 'greatest blessing of the age'. He also boasted it was 'as delicate as a banknote', although only the wealthy were in a position to confirm it. The modest claim of a rival firm was that their product was 'splinter free'.

In England toilet paper was affectionately known as

'bumph' or 'bum fodder'. In my youth soft, absorbent tissue in pastel colours was far away in the future. Instead we used Izal off-white, defiantly non-absorbent tracing paper smelling of disinfectant. It may have been this that inspired many people, including my grandmother, to tear newspaper into square sheets and thread them on string to hang up in the loo. They worked fine: the printer's ink acted as a disinfectant and you could sit back and catch up on yesterday's news.

My first book, *Stars Beneath the Sea*, which I thought was a history of pioneering deep-sea divers, was praised as a 'Book of the Month' by a reviewer who categorised it as being about 'sanitary engineering'.

INSIDE INFORMATION

'X-rays will prove to be a hoax'
– Lord Kelvin, President of the Royal Society, 1896

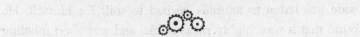

Surgery was an academic abattoir. The operating theatre was an auditorium in which surgeons displayed their bravura butchery. Robert Liston, a nineteenth-century Scottish surgeon, boasted he could lop off a leg in twenty-eight seconds. He began each performance with the words: 'Time me gentlemen, time me.' In his haste he once detached the patient's leg and a testicle plus two fingers from his assistant. Speed was essential if the patient was to survive the shock from pain and blood loss. One surgeon thrust his walking stick across the patient's mouth and told him to bite down. The stick with its deep teeth marks survives, which is more than most of the patients did.

Sweet dreams

Surgeons had searched for something that would put the patient into a stupor, but they had turned to cocktails of poisonous herbs that did indeed put the patient into a deep and sometimes permanent sleep. They also tried opium – and kept some for themselves. Some surgeons bled the patients until they passed out or put a wooden bowl on their head and whacked it with a mallet.

Scientists were testing the properties of new compounds, including volatile gases. Humphry Davy was a great sniffer of gases and his colleagues were always relieved when he turned up for work the next day. When he heard that nitrous oxide was lethal to animals he had to sniff for himself. He found that it was 'highly pleasurable' and was soon inhaling it three or four times a day. Davy gave it the jolly name of 'laughing gas' and it soon became all the rage at parties. Medical students, as ever, loved to intoxicate themselves.

Davy noted that when he sniffed the gas the pain from an inflamed gum was immediately quenched and he wrote that as it 'appears capable of destroying physical pain, it may probably be used with advantage during surgical operations'. What a discovery, but Davy, one of the most brilliant minds of his time, did nothing about it.

Laughing gas became a fairground favourite and in 1844 'Professor' Gardner Colton brought his gas-fuelled frolics to the entertainment-starved folk of Hartford, Connecticut. His advertisement stated that he was bringing 'forty gallons of Nitrous oxide Exhilarating or Laughing gas . . . administered to all in the audience who desire to inhale it'. Those who did were guaranteed to 'Laugh, Sing and Dance, Speak or Fight'. And they did.

In the audience was a local dentist, Horace Wells, who noticed that one of the volunteers had danced so vigorously that he had gashed his leg on a table, yet seemed to be unaware of the wound. Wells invited Colton to his house the next day and got him to administer the gas to him while a fellow dentist extracted a tooth. It was entirely painless. In those days extractions were excruciating, so toothache was usually dulled just by the sight of a dentist and his blood-stained pliers. Wells envisaged 'a new era in tooth pulling'.

After fifteen more painless extractions he decided he would announce this miracle to the world and arranged a demonstration at the Massachusetts General Hospital in Boston. There was a large audience of medical students and doctors, among them John C. Warren, the most distinguished surgeon in America. Unfortunately, Wells did not give sufficient gas to the patient. When the tooth was pulled the patient cried out in pain. To hisses and boos, Wells left the stage, a broken man who had fluffed his big chance. He sold his house, abandoned dentistry and became addicted to nitrous oxide, but it failed to make him laugh.

A former partner of Wells also became interested in drugs that caused insensibility. William Morton was a go-getter and could see a golden future in putting people to sleep. Ether had become the latest craze and there were communal ether 'frolics' for 'doctors and women of refinement'. To see if it was safe as an anaesthetic, Morton practised on his wife's pet dog and her goldfish. They survived, so he put an ether-soaked cloth over his face and immediately collapsed. What he didn't know was that such a dose could prove fatal. Luckily, the cloth fell from his face and he recovered.

After he had carried out 160 painless tooth extractions Morton felt confident to go public. He chose the Massachusetts Hospital, where the same sceptical Dr Warren was to operate. He cut out a cyst the size of a billiard ball from the patient's neck in complete silence. At the end there was prolonged applause and Warren announced: 'Gentlemen, this is no humbug.'

Within days the word about painless surgery was known around the world and Morton was selling the anaesthetic and a dispenser that he had invented in two days. Oliver Wendell Holmes suggested the word 'anaesthesia', which had been coined by Dioscorides almost two thousand years ago. Morton patented his invention and was selling shares. His product was called Letheon, Virgil's name for opium-induced sleep. There was no mention of ether because it had been synthesised in 1540. If it leaked out that it was only ether his claims would collapse.

Charles Jackson, a chemist who had suggested ether to Morton, asserted it was *he* who invented anaesthesia. To avoid expensive litigation he was granted ten per cent of the profits. Then another claimant, Crawford Long, produced affidavits to prove that he had carried out several operations under anaesthetic even before Wells. Various institutions supported one or other of the claimants, but although it was obvious that Long was first, he had not written an article or brought his activities to the attention of the medical fraternity. However, it was Morton who launched anaesthesia on the world. Jackson was seen jumping on Morton's grave because the headstone stated that he was the inventor of painless surgery.

Many surgeons grasped anaesthesia as a gift from the gods, the end of agonising operations. A surgeon said after his first day of operating with ether: 'I realised that on that day I was not writing history, but making it.' Yet in Britain and elsewhere many patients were still in agony on the operating table twenty years after Morton's demonstration. They feared the unpredictability of ether. Sometimes the patient hardly closed their eyes or awoke prematurely, while others were deeply comatose.

No one knew what the safe dose was. The major cause of uncertainty was that vapour was administered on a cloth or sponge held over the nose and mouth, which made it impossible to judge the concentration.

John Snow, a general practitioner in London, was determined to do something about it. He had already produced a pump to resuscitate stillborn babies. Then he studied ether and found that the concentration of ether vapour was dependent on temperature, so he experimented. What discomforted his landlady was not just the smell of ether in every room of the house, but the explosions that rattled her display cabinet. He invented an inhaler with a vapour spray controlled by changes in temperature facilitated by the addition of warm or cold water. It also had a close-fitting mouth and nose cup with taps to make a gentle transition from air to ether. 'Lightning' Liston, the best surgeon in Britain, was impressed and hired Snow. Within a couple of months Snow had transformed ether use throughout London and made surgery safer. He gave meticulous instructions on how to maintain a safe level of anaesthetic and invented the role of the anaesthetist.

At about this time Liston's protégé James Simpson, Professor of Midwifery at Edinburgh University, was searching for a gas that might ease the mother's pain during childbirth. He held regular parties at his home at which he and colleagues sniffed whatever volatile substance they had come across. The samples tested included acetone, now best known as nail polish remover, ethyl nitrate, a constituent of rocket fuel, and benzene, a poison and potent carcinogen. One night they sniffed the vapour of a sweet-tasting liquid. Fifteen minutes later they woke up under the table – the bottle contained chloroform. Four days later Simpson gave chloroform to a woman with a history of difficult births. She came to unaware that she had delivered. Her baby daughter was christened Anaesthesia. By the end of the 1840s chloroform had displaced ether.

Snow also dispensed chloroform because it was sweet-tasting, whereas ether was harsh to inhale. I had an operation when I was nine and sixty-four years later I can still recall the taste of the ether. But the range of safe concentration of chloroform was much narrower than that of ether. With a third of a teaspoon the patient was still conscious; half a teaspoon could be fatal. Snow's inhaler was essential to avoid overdosing. In his career of 4,500 operations and births, he had only one fatality. But when the procedure was in the hands of untrained general practitioners who shunned the inhaler in favour of the good old handkerchief or sponge soaked in anaesthetic, the death rate was high. Apparently there was also the possibility that after the operation a female patient didn't just recover, she became a nymphomaniac.

Some surgeons went back to ether, with which the death rate was lower. The Chief Medical Officer for the British Army disapproved of *all* anaesthetics for battlefield amputations. He was convinced that it was 'much better to hear a man bawl lustily than to see him sink silently into the grave'. Luckily, his surgeons just ignored his orders, anaesthetised their patients and got on with 'doing nothing but cutting off arms and legs all day long'.

After six 'wretched' pregnancies Queen Victoria opted for chloroform for her seventh delivery. Snow managed the anaesthesia and the Queen found it to be 'delightful beyond measure'. *The Lancet* growled that any medical man who would risk the Queen's life with a poisonous gas was a bounder. The Queen and Prince Albert did not agree and Victoria used chloroform again when her last child was born.

Snow had shown that in skilled hands anaesthetics were safe, but there were very few specialist anaesthetists who knew what they were doing. Before the Great War there were no more than ten in the whole of the United States and Britain was no better. In 1937 Robert Macintosh was appointed Britain's first-ever Professor of Anaesthetics. Before he took up residence at Oxford University he spent months learning about anaesthetics, including a stint at a field hospital during the Spanish Civil War. He went on to invent an improved version of Snow's portable inhaler and instigated refresher courses for military medics. By the time the Second World War was over, anaesthesia was a respectable arm of modern medicine.

Snow, the first professional anaesthetist, also discovered that cholera came from drinking contaminated water. When he died his friends had to club together to buy a headstone for his grave. There is also a pub in London's West End that bears his name.

With anaesthesia and aseptic surgery, parts of the body such as the chest, abdomen and even the brain were now available to the surgeon. But a physician must try to diagnose what is wrong with the patient before he recommends an appropriate treatment. Ideally he should do so before the patient is cut open.

Listening in

For the overworked doctor the reward after a long day was to press his ear to a lady's bosom to listen to her heart or just daydream. The ancient Greeks did their share of chest listening and heard sounds such as 'boiling vinegar' and 'creaking like new leather'. What could it all mean? William Harvey and Robert Hooke heard the rhythm of the heart. Hooke even speculated: 'Who knows, I say, that it might be possible to discover the motions of the internal parts of bodies . . . by the sound they make . . . and thereby discover what . . . is out of order.'

His prophecy came true in France at the beginning of the nineteenth century. René Laënnec was a brilliant and ambitious young doctor but he was in a quandary. He needed to listen to the chest of a voluptuous young lady. Thinking his ear might be inappropriate, he rolled a sheet of paper into a tube and placed his ear and her breast at different ends of the tube. He was amazed at how clearly he heard her rapidly pounding heart and excited breathing. He spent the evening on a lathe, turning a hollow wooden tube 1 ft (30 cm) long and slightly flared at one end. He called his invention the stethoscope (Greek for 'chest observer').

The chest cavity was an auditorium of sounds and as Laënnec was a musician he could describe each sound accurately. He listened to the chests of seriously ill patients and then examined them at the post-mortem to determine what lesions they had. Thus he correlated the sounds to actual diseases. He could distinguish pneumonia, bronchitis, pulmonary tuberculosis, emphysema, pulmonary infarction and pleurisy.

Laënnec wrote a nine-hundred-page guidebook on the use of the stethoscope and how to recognise the sound menu of the chest, heralded as 'the most complete treatise on the diseases of the chest'. For a small charge, a stethoscope was included and many doctors gave it a try.

There were critics, of course, who claimed it was quackery or an affectation, and many doctors were too deaf to distinguish the sounds and too blind to accept technology that bypassed their wisdom. François Broussais dismissed the book as 'a collection of undisputed facts or useless discoveries'. He believed that the solution to everything was to bleed the patient. It is estimated that he shed more blood than all the wars that occurred during his lifetime.

Laënnec could diagnose everyone's condition except his own. What he insisted was asthma was tuberculosis, which in those days was the greatest killer of humanity. He was a fine lecturer and students and doctors came to Paris from all over the world to hear what he had to say. He was a great pathologist and was the first to describe cirrhosis of the liver and peritonitis. He was made a Knight of the Légion d'Honneur.

Laënnec moved diagnosis into a new era in which the physician would rely less on what the patient told him or on his personal bias, and instead could use objective evidence. The modern stethoscope with two earpieces and a small cup dangling on a tube is indispensable. It is the doctor's badge of office. Had it not been invented, what on earth would medical students hang around their neck to give the impression they know what their doing?

Bones on view

What the stethoscope had been to the nineteenth century, X-rays became in the twentieth century. Their discoverer was a German, Wilhelm Röntgen. As a boy he was a star pupil until he drew an unflattering caricature of his teacher on the blackboard and was expelled. This didn't seem to retard his education because he became Professor of Physics at the University of Würzburg.

In 1895 Röntgen was busy experimenting with cathode rays in his laboratory. In the darkened room he noticed that a chemically treated paper on the other side of the room was fluorescing. It couldn't be the cathode rays because they couldn't reach that far and anyway he had encased the cathode-ray tube so that no rays escaped. He immediately grasped that it was the work of a previously unknown, invisible ray.

He attempted to block the waves with several sheets of paper held between the tube and the paper screen, but they did not appear as a shadow, nor did two packs of cards, a thousand-page book or several sheets of tinfoil. Finally he tried a lead weight and this did leave a dark shadow on the paper, proving that it had absorbed the rays, but to his horror the lead weight was held up by *a skeleton's hand*. It was unreal and it was impossible.

Röntgen didn't return to the lab for several days, until the Christmas break, when no one would be around. He put his hand in the path of the rays and there on the paper was its faint shadow, with the bones beneath boldly displayed. Clearly these powerful rays could pass through flesh but not dense bone. He called them X-rays after the mathematical X, meaning the unknown.

He persuaded his wife to have her hand photographed by the rays. The picture shows the skeleton perfectly and it's wearing two rings. Röntgen sent copies of the photograph and a letter explaining his discovery to several physicists and also his father, who was a newspaper editor in Vienna. He published the story and within a week it was headline news all around the world.

The public panicked, but not at being reduced to a skeleton – indeed X-ray photo studios did a brisk business – but that rays might look through their clothing and reveal what lay beneath. One enterprising company brought out a new line in X-ray-proof knickers and a ditty of the day warned:

> *'I hear they'll gaze*
> *Through cloak and gown and stays,*
> *These naughty, naughty Roentgen rays.'*

The State of New Jersey proposed a law to ban X-rays in opera glasses.

Within months X-ray machines were being manufactured and used to look for fractured bones, gallstones and bullets that had somehow found their way into the human body. In 1897 Walter Cannon, a brilliant young American physiologist, gave animals capsules of bismuth in their feed and found that they were opaque to X-rays and therefore could be used to observe the stomach and gut in action. Later it was found that barium was better for internal X-rays and the Horlicks Company made the barium meal.

Within a year of Röntgen's discovery a thousand scientific articles on X-rays had been published. He could have made a fortune but refused to patent his invention – it was for humanity. His reward was to become the recipient in 1901 of the very first Nobel Prize for Physics.

In a court case a judge refused to accept X-ray photos in evidence as it was 'like offering the photograph of a ghost'. Thomas Edison predicted that there would soon be an X-ray machine in every home, but didn't say what for. When I was a child there was a 'pedoscope' in all the better shoe shops that X-rayed your feet to ensure the shoe was a good fit.

Kids loved to look through the viewer to see their skeletal toes wriggling. I wonder if they've still got all those toes?

Röntgen's invention was a two-edged sword. It was a powerful diagnostic tool, but as the early radiologists found to their cost, over-exposure to the rays was dangerous, as I am reminded every time I have an X-ray of my teeth and the dentist scurries out of the room. X-rays burn the skin, yet carefully used they remove warts and treat acne and *Lupus vulgaris* (skin tuberculosis). They can cause cancer yet they are the basis of radiotherapy to kill cancer. Had Röntgen not discovered X-rays Henri Becquerel would not have been stimulated into discovering radioactivity. Without the X-ray diffraction of crystals James Watson and Francis Crick would not have determined the structure of DNA.

Safe for babies

Because X-rays were dangerous, in the late 1950s Ian Donald, Professor of Midwifery at Glasgow University, adapted submarine-detecting sonar into an instrument to detect unborn babies. He was not the first to play with ultrasound to look inside the human body. Other scientists found that the skin was a barrier to ultrasound, so they immersed the patient in water. Picture a patient sitting on the bottom half of a dentist's chair. Around his waist is a child's inflatable paddling pool with rubber pantaloons affixed to the floor of the pool to accommodate his legs. The pool is filled with water which continually swills over on to the floor.

Professor Donald solved the problem by simply smearing oil on the skin of a pregnant woman and passing a transducer across her abdomen. A computer enhanced the echoes from

Inside Information

the foetus and displayed them on a screen. At first the images
were clear only to an experienced operator. My wife gave
birth to both our children in Professor Donald's hospital and
when I saw the scan pictures I had no idea what she was
incubating. Ultrasound scanning was quick and caused no
distress to the foetus. It became the routine for monitoring
the development of the foetus. And now parents take home
a detailed portrait of their unborn baby.

It was inevitable that someone would come up with a
telescope to look inside the body's various orifices. Originally
it was a straight metal tube not unlike a submarine's periscope
and to the patient it felt like it. If the passage it entered had
a bend in it the physician simply heaved to starboard and
the patient's eyes watered. It's called rigid endoscopy and it
means what it says. With the advent of fibre optics along
came the flexible endoscope that could slide round corners.
In 1965 Professor Harold Hopkins at Reading University
used wafer-thin glass fibre as lenses that delivered light and
relayed the view back to the observer. Now such examinations are routine and barium meals are a thing of the past.

Scanning new terrain

The most significant advance since X-rays is computer-aided
tomography, a technique that involves passing X-rays through
the body to produce images of cross-sections of internal tissues
or organs without shedding a drop of blood. The 'slices'
recorded in this way can be viewed on screen individually
and the computer can integrate them into a three-dimensional
image.

The inventor was Godfrey Hounsfield, who worked for

Electrical and Musical Industries. In the late 1960s EMI was accumulating so much money from the Beatles' records that it branched out into high-tech medical equipment. Hounsfield's CAT (Computerised Axial Tomography) scanner could look into the human body in extraordinary detail. Indeed it produced such an enormous amount of information that he had to programme the computer to make an educated guess at the result and then modify it as more data comes in. This ingenious procedure is now widely used for coping with vast amounts of data.

Initially the problem was the slow speed of existing computers. It took four hours to scan the organ and convert it into an image. As computing power increased Hounsfield produced a machine to scan the brain with an X-ray beam circling around the head. In 1975 he produced a whole-body scanner. The patient lay on a platform that moved through the central hole in the doughnut-shaped machine in which the X-ray emitter rotates around the body. Hounsfield was awarded the Nobel Prize for Medicine in 1979.

The imaging technology of the CAT scanner led to other scanning machines that didn't rely on X-rays. The PETT (Positron Emission Transaxial Tomography) scanner was designed to look at the brain. After a stroke, parts of the brain can be damaged and the victim may lose functions such as speech or limb mobility.

In 1984 Louis Sokoloff invented a scanner that could 'see' the brain at work. He injected a patient with slightly radio-active glucose that accumulated in the glucose-hungry brain in the regions of greatest activity. These areas give off energy in the form of gamma rays that can be detected by probes on the patient's head and converted into a map of the brain. Brain activity could be viewed when the patient was consciously remembering or planning, for example, and any dysfunctional areas were highlighted.

The most sophisticated scanner relies on magnetism. In a powerful fluctuating magnetic field many atoms, such as hydrogen, resonate and align with the magnetic field. When the magnet is off, the atoms 'relax' back into their former orientation. Raymond Damadian, an American doctor, noticed that cancerous cells were much slower to relax than healthy ones and these changes could be transformed into 3D images. In the scanner the patient lies inside a tunnel surrounded from head to foot by a spiral magnet and exposed to a fluctuating magnetic field. Another American, Paul Lauterbur, and Peter Mansfield, a Briton, developed the techniques that made cancer diagnosis possible. Magnetic Resonance Imaging (MRI) could detect tumours and soft-tissue damage as well as monitoring muscular dystrophy and transplants. Its images are ten to thirty times more detailed than those of X-rays. We have come a long, long way.

BLOODY HELL

'Bloody instructions which . . . return
to plague the inventor'
– Shakespeare

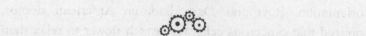

Whatever the means of diagnosis, surgeons usually resort to cutting. Their greatest enemy was blood. A young man of the average adult male weight of about 155 lb (70 kg) has 10.5 pints (6 litres) of blood. If a surgeon accidentally nicks an artery, blood can explode out in an unstoppable gush. With a major injury, within three minutes half the body's blood can be lost and with it the patient.

How could the flow be staunched? Ambroise Paré was still prescribing ancient cures such as puppies boiled in oil and 2 lb (0.9 kg) of worms drowned in white wine, but this Frenchman was the greatest surgeon of the Renaissance. He learned all about blood loss when treating battlefield wounds. In his day the practice was to cauterise the wound with a red-hot iron that melted tissue and sealed torn blood vessels. A bullet wound was treated by pouring boiling oil into the hole. Such desperate measures did enormous damage and were excruciating for the patient.

Paré shunned such methods and concentrated on closing damaged blood vessels. He invented a clamp that would slow the blood loss while he ligated (tied off) the artery with silk thread. In practice it was difficult to fix a clamp on a slippery artery and it often took several ligatures to halt the flow.

The tourniquet was a simple device known to the ancient Egyptians and was used in amputations. If a bandage was tied tightly around the limb above the wound it restricted the blood flow lower down. Obviously the blood supply could only be curtailed for a short time.

It's sew simple

A major advance came at the dawn of the twentieth century when a young French surgeon took embroidery lessons. Alexis Carrel at Lyon Hospital learned to sew tiny stitches and he ingeniously gathered the open end of a blood vessel in three places so that when he pulled all the threads simultaneously it opened up into a triangle. This made it far easier to attach the two ends together again. However, it caused clotting. The stitches were an obstacle in the otherwise smooth and slippery interior of the artery. Blood cells built up on the stitches and if this clot broke free, it could travel to the heart or brain, causing a catastrophic blockage. Carrel's solution was to roll back the edges of each of the severed ends to form a collar. The collars could then be stitched together with no stitches penetrating into the tube. His techniques transformed vascular surgery.

Transfusions

Blood loss would be less serious if it could be rapidly replaced. As early as 1650 an Englishman, Francis Potter, tried a direct transfusion using a sharpened quill to pierce the patient's vein and tubes made from a sheep's windpipe. It failed. In 1666 Christopher Wren, the architect of St Paul's Cathedral, exchanged blood between dogs. The following year Frenchman Jean-Baptiste Denys conducted a demonstration at the Royal Society in London in which he transferred blood from a sheep to a man who was 'a little cracked in the head' otherwise he wouldn't have volunteered. Apparently he survived, but goodness knows how. Many people feared the transfusion made the patient acquire the sheep's characteristics. In his play of 1676, *The Virtuoso*, Thomas Shadwell envisaged a deranged scientist creating an entire flock of sheep-men, each with 'wool growing on in great abundance', and a 'sheep's tail'. Transfusions from animals to humans were banned.

By the nineteenth century several doctors were transfusing blood from one person to another who was ill. Dr Blundell of London claimed that his transfusions had never been fatal, which is difficult to believe as several countries had banned transfusions because of the high death rate. A pope on his deathbed had blood transfusions from three youths. They all went to heaven together.

It was puzzling that some transfusions were fine while others led to a rapid death. This inconsistency was not explained until 1901 when Austrian pathologist Karl Landsteiner experimented on blood. He and his colleagues donated drops of blood and these were mixed together in pairs. In some mixes the blood coagulated, whereas in other pairings there was no reaction at all. Obviously there were different 'types' of blood, which Landsteiner labelled A, B, O and AB, some of which were incompatible with each other. He originally identified four blood groups and two other proteins that were responsible for the immune reactions and he worked out which combinations were compatible for safe transfusion.

Landsteiner's transfusions were made directly from person to person because blood clotted if stored. In 1914 sodium citrate was added to stored blood to prevent clotting. Blood banks were introduced during the Second World War and for the first time military surgeons had as much blood as they needed.

My blood group is O-negative, which is found in less than eight per cent of Caucasians. It has none of the antigens that cause the other blood groups to react against each other. Therefore O-type blood can be given safely to anyone whatever their blood group.

In 1981 American biochemist Jack Goldstein extracted an enzyme from green coffee beans that could snip off the antigens from red blood cells to change these to O-type, but it could not be made in sufficient quantities to be commercially viable. In 2007 Henrik Clausen at the University of Copenhagen used enzymes from fungi to convert both A- and B-group blood cells into O-type.

For several years there have been concerted efforts to produce artificial blood for transfusion. Unfortunately, trials have shown that patients given artificial blood did less well than those given saline solution. It seems that the answer may lie with embryonic T-cells that can grow into red blood cells. It is predicted that artificial blood will be a reality within a decade.

Aid for ailing organs

Our organs strive ceaselessly on our behalf and we don't even notice, at least not until they malfunction. Often there was nothing a doctor could do for a failing organ. Willem Kolff, a Dutch doctor in Leiden, was dismayed watching his patients slowly dying as inflammation of the kidneys caused the accumulation of waste in the body. He determined that if a man could grow an organ, he could build one.

He wasn't the first to try. As early as 1913 researchers at Johns Hopkins University in the United States were experimenting on dogs with the aim of producing an artificial kidney. They had problems with blood clotting, so they ground up the heads of thousands of leeches because they contained an anti-coagulant that kept the victim's blood flowing while they fed on it. The outbreak of war in 1914 curtailed their research – much to the leeches' relief.

The kidneys regulate the salt and liquid content of the body and eliminate waste products as urine. A kidney contains a million or so filters able to process 50 gallons (227 litres) a day. Replicating such a complex organ would be difficult. Kolff started with a sausage. Thomas Graham, a Scottish chemist, had shown in the nineteenth century that liquids could diffuse through a permeable membrane. Kolff found that his butcher's sausage 'skins' were just such a membrane.

The patient's blood would be diverted into the permeable tubing, the longest meatless sausage in the world, which was wrapped around a rotating drum half-submerged in a bath of salty water. The waste passed into the salty water and the refreshed blood was pumped back into the patient. It took many years to perfect and was built from bits and pieces that he had scavenged. Kolff was also diverted by helping Jews to escape from the Nazi-occupied Netherlands and sheltering Resistance fighters. Ironically, the first of his patients to undergo a successful dialysis was a Nazi sympathiser. On its completion she opened her eyes and declared she was going to divorce her husband.

Kolff's dialysis machine was bigger than a fat man, but it worked. In theory patients could survive indefinitely without functioning kidneys, although even as late as 1970 such a patient needed to be on dialysis for thirty hours a week. The limiting factor was that for each transfusion a glass tube had to be inserted into a vein and the same vein couldn't be used twice, so the patient rapidly ran out of accessible veins. In the 1960s patients were fitted with a shunt, a permanent connection to a vein to which the dialysis tube could be attached any number of times.

Kolff never patented his machine, and indeed he sent them to researchers in several countries in order to alert hospitals to what they could do, and perhaps stimulate others to produce other artificial organs that worked outside the body.

In 1896 a British surgeon had suggested that 'no new method and no new discovery can overcome the natural difficulties that attend a wound to the heart'. But with anaesthesia, aseptic conditions and improved techniques to control bleeding, surgeons became emboldened. It had been almost universally accepted that certain parts of the body, such as the brain and the heart, were off limits to the scalpel and would always be so. An influential surgeon even warned that any surgeon 'who tries to suture a heart wound would deserve to lose the respect of his colleagues'. But some surgeons would not be shackled.

In 1883 William Halsted, a young American surgeon, saved a patient dying from carbon monoxide poisoning. The gas attaches itself to the red blood cells and prevents life-giving oxygen from doing so. Halsted drained the blood from the patient's arm into a jar. He stirred it to expose it to the air and prevent clotting and then returned it to the patient, thus saving his life. This principle would make possible great advances in heart surgery.

The problem was that the heart is the most active organ in the body, so it has to be stopped if any serious repair work is needed. But the rest of the body relies on the heart to circulate oxygen, as without it the tissues die. Walter Lillehei of the University of Minnesota found a way round this dilemma. All that was needed was a fit volunteer with a blood group compatible with that of the patient. The volunteer's blood would be pumped into the patient's body and his lungs would aerate and pump the blood for them both. Meanwhile the patient's heart was clamped off and the surgery could proceed. It worked and both hearts and bodies survived.

There followed several operations on seriously ill small children with a parent providing the heart power to keep the child alive for the twenty minutes or so of surgery. Sixty per cent of his operations on children with multiple heart defects were successful and saved lives. Lillehei became hot news, but it was controversial because the procedure was extremely risky. It was the only operation that had a chance of a 200 per cent mortality rate, so no other surgeons copied his procedure.

In 1931 American surgeon John Gibbon wondered whether a machine could temporarily take over the job of aerating and circulating the blood while the heart was clamped off and inert. He decided to invent one. If he had known it would take twenty years, he might have had second thoughts.

He encountered numerous problems. The machine was huge and required buckets of blood to get it going. Substantial leaks were frequent and he was often paddling in blood – not the ideal floor covering for an operating theatre. However, in 1953 he mended a hole in the heart of a young woman called Cecelia. For almost half an hour she was entirely dependent on his heart and lung machine.

The greatest challenge was how to oxygenate the blood as efficiently as the lungs. The human lung is a labyrinth of damp tunnels with the surface area of a tennis court. How could a machine provide such an enormous oxygen-absorbing area in an operating theatre? Having a flat surface would require a spacious gymnasium. Other surgeons had used the lungs of monkeys or dogs suspended in oxygen-rich jars as surrogate lungs attached to the patient's bloodstream. Surprisingly, Gibbon never thought of monkey lungs. It was just as well, for more patients died than survived.

The most effective way of oxygenating blood is to bubble oxygen through it. However, if even a single small bubble got into the patient's bloodstream it could cause a fatal embolism. Lillehei's protégé Dick DeWall cracked the problem. He added a chemical used for smoothing mayonnaise, which disrupted the skin of bubbles, and then he passed the blood through a descending spiral tube. The denser, bubble-free blood ran down the tube, whereas the more buoyant blood with a bubble rose upwards. The Lillehei–DeWall Bubble Oxygenator was the first relatively cheap, simple and reliable heart and lung machine.

A critical moment was when reawakening the heart. In the eighteenth century Benjamin Franklin had suggested that electricity might revivify the recent dead. John Hunter, the greatest surgeon of his time, recommended restarting a stilled heart by stimulating it with electricity. In 1774 he successfully revived a three-year-old girl who had fallen from an upstairs window by giving electric shocks to her chest to restart her heart. Defibrillation is now commonplace.

These inventions also revived not just patients but the surgeons' dream that at last every organ in the body was at their mercy.

HEARTACHES

'Transplantation of organs such as the kidney,
I have never found positive results to persist'
– Emil Theodor Kocher, Swiss surgeon

Transplanting body parts to replace failing ones is not a recent idea. The eighteenth century was not a good time for teeth. New plantations in the Caribbean exported sugar to Britain and it proved to be as addictive as opium. The teeth of the wealthy rotted into blackened stumps. A lady's smile revealed a miniature neglected graveyard that detracted somewhat from her charms.

Dentures had been manufactured by the Etruscans. A 2,700-year-old human skull had false teeth carved from bone clamped between two strips of gold with spaces left to fit over the remaining teeth. It looks very uncomfortable to wear.

By the seventeenth century there were improved dentures, but the upper ones kept falling out. Some vain ladies had their gums pierced with hooks to anchor the uppers. A French dentist provided a spring to keep the upper plate in place, but it was so strong that wearers couldn't close their mouth. By the time he became president George Washington had only one tooth. His spring-loaded dentures carved from hippopotamus ivory were never satisfactory. His teeth failed to look as good as they had in the mouth of the hippo. Surely there must be a better alternative.

Teeth for sale, one careless owner

As early as the mid-sixteenth century Ambroise Paré replaced a rich woman's decayed incisor with a new one 'volunteered' by her lady-in-waiting. Within a hundred years dentists had a lucrative practice relocating teeth no longer needed by their previous owner. There was a ready supply of 'human ivory' from mortuaries, dissecting rooms and battlefields. It was an offence to extract incisors from live men of military age because healthy front teeth were needed. Grenadiers had to bite open the fuse on their grenades and musketeers yanked the wooden caps from their cartridges with their incisors.

There was, however, no embargo on removing teeth from dead men. Scavengers searched the battlefields for a healthy victim, or at least one as healthy as a corpse can be. So-called 'Waterloo teeth' were on offer forty-five years after the battle. Teeth from those fallen in the American Civil War were shipped to England to adorn the mouths of toothless toffs.

James Spence, a London dentist in the mid-eighteenth century, specialised in high-class ladies who demanded superior teeth that had not spent a lifetime chewing tobacco or grinning at whores. When Spence advertised for young maidens with good teeth for sale, it brought a constant queue to his door. His procedure was to choose the healthiest-looking mouth and extract two or three teeth to ensure at least one would be a good fit in the recipient's gap. He hurried into the next room to remove the client's rotting stump, thrust the donor's tooth in the bloody socket and tied it to the adjacent teeth with silken thread.

One of Spence's potential donors chickened out at the last moment. Had she donated her two front teeth it is unlikely she would have become Lady Hamilton or Admiral Lord Nelson's mistress. Nelson, however, had no right to be fussy about a girl's teeth: his were so bad that he never smiled to reveal them.

The poor donors were given a small payment, whereas the client paid a generous sum. New teeth became a must-have improvement much as breast enlargement is today, and just as expensive. Some saw the practice as robbing the poor to enhance the rich. The King's dentist was one of the few medical men to condemn it as 'ineffectual, dangerous and . . . expensive'. He was right. Most of the implants failed, although a few were said to have lasted for several years, and there was the risk of transferring a disease from the blood on the implanted tooth. There were cases of syphilis that were probably contracted in this way.

When ceramic teeth and moulded denture plates became available, human tooth implants fell out of favour. For a while denture plates were made from celluloid, which was cheap, easy to clean and inflammable. It also melted when challenged by hot tea and after a man's dentures caught fire while he was smoking, celluloid was replaced by plastic acrylic resin. Implants are back in fashion: they deploy a ceramic 'tooth' that fits over a titanium rod screwed into the jaw.

Drilling and filling joined the dentist's skill set. The much-loved dentist's drill was invented by the Rome-based Greek surgeon Archigenes. It was a clumsy device powered by a rope. The drilling was slow and therefore excruciatingly painful. The modern air turbine spins at 200,000 revolutions per minute, but its threatening hum retains the fear factor.

It took a long time to convince the British public that frequent brushing would maintain their teeth. In 1970 dentists in Britain carried out thirty million fillings and four million extractions. I lived in Glasgow in the 1970s and my wife had an operation there. While being trolleyed to the theatre she was frequently told to take out her dentures and put them in the pot provided. The medical staff were astonished that a thirty-three-year-old woman didn't have removable teeth. It had been a Scottish tradition for a bride to have all her teeth extracted to avoid subsequent inconvenience and expenditure. A man often had his teeth removed for his twenty-first birthday.

New parts for old

Unlike teeth, our bones are continuously replacing themselves. Our entire skeleton is renewed every ten years or so. Its powers of regeneration encouraged surgeons to attempt to repair or replace worn joints with artificial ones.

Prosthetics are not new. Herodotus recounts the story of a criminal awaiting execution who sawed off his foot to escape from shackles and hopped away to freedom. He carved a wooden foot and although the authorities were searching for a monopode he was caught and hung.

Iron hooks and wooden peg legs were replacements for lost hands and legs many centuries before they became fashionable for pirates. They stayed in vogue for a long time. In the 1960s I met a fisherman in the Canary Islands who had a peg leg, having lost a limb to a shark, or so he said.

In the sixteenth century Ambroise Paré invented numerous prosthetics, including artificial limbs, glass eyes, and false noses for syphilis sufferers. His mechanical hand with gears and movable joints was astonishing. As early as 1508 a German knight named Götz von Berlichingen replaced his missing arm with a sophisticated metal limb with hinged and sprung joints.

Building a replacement for the weight-bearing ball-and-socket joint of the hip was more difficult than inventors thought. They tested a variety of materials but none slid smoothly in the socket and the ball-shaped bit that was driven into the sawn-off top of the thigh bone sometimes fell out. It was not until 1960 that John Charnley, an orthopaedic surgeon at Manchester Royal Infirmary, made a successful hip replacement. He used a highly polished metal ball in a socket lined with high-density polythene that was as strong and generated as little friction as the natural joint. He manufactured his first hip cups in his shed at home.

As usual there were teething problems. My PhD supervisor had a hip joint replacement in the early 1960s. She later suffered sharp pains in her hip. The implant had been made of two different metals and they generated electricity like a battery. Nowadays the globular heads are made of ceramic.

Transplanting soft-tissue organs inside the body had unforeseen problems. The obvious organ to begin with was the kidney because it had few arteries and veins and was therefore relatively easy to detach and reattach. The first attempt was in 1909 when French surgeons replaced kidneys with organs taken from animals without success.

In 1947 Charles Hufnagel, a Boston surgeon, successfully transplanted a human kidney into a patient with faltering kidneys. She was so ill she couldn't be taken to theatre, so Hufnagel brought the kidney to her bed and operated by the light of two desk lamps. As soon as the kidney was connected to her blood supply it began to work. After a few days the implanted kidney started to fail, but by then her blood was cleansed and she was restored to health.

Four years later two French surgeons, Charles Dubost and Marcel Servelle, attended an execution. They waited patiently until the blade of the guillotine separated the felon's head from his body. Then they rushed the decapitated corpse to a table, whisked out his still-warm kidneys and transplanted them into two women with diseased kidneys – one organ each. The operation went well and the women rallied, but within nineteen days both of them died. The post-mortems revealed that the implanted kidneys had shrivelled and degenerated in just a few days. Why?

Rejection

Alexis Carrel had carried out numerous bizarre organ transplants between laboratory animals and discovered that transplants from the same animal usually 'took', but those between two different animals of the same species were invariably fatal. English biologist Peter Medawar had been experimenting with tissue grafting with a view to patching up wounded soldiers. He attributed the failure of his grafts between different animals to the recipient's immune system. The implants weren't merely failing: they were being rejected.

The body's immune system is always on a war footing, alert to any intrusion. As soon as 'foreign' material is detected it is attacked. An implant of the body's own tissue is recognised as being 'self' and therefore the immune response is not triggered. This had been demonstrated in 1954 when American surgeons Joseph Murray and Hartwell Harrison made the first successful kidney transplant. It was between identical twins. They had the same genes and blood group, even shared the same placenta in the womb. So the body of the recipient considered his brother's kidney to be 'self', not 'alien'.

The heart was surgeons' Holy Grail. It is the hardest-working organ in the body, contracting over 100,000 times a day, so perhaps it can be forgiven when it fails. Forty per cent of us die from cardiovascular disease. A heart attack can be mercifully sudden. One man was found on a park bench crouched over, with his hands still holding the shoelaces he had been tying. In contrast a gradually failing heart is a prolonged nightmare.

At first, surgeons tried to repair worn-out parts. Heart–lung machines enabled the heart to be stilled long enough for surgery. In 1952 Charles Hufnagel inserted an artificial valve into a patient's aorta. The device he used was almost identical to one patented in 1858 – as a bottle stopper. The job of the valve was to ensure that the blood flowed only in one direction. Hufnagel's prosthetic was a cylindrical cage with a plastic tube at one end containing a marble. A heartbeat pushed the marble into the cage, which allowed blood to flow around it. Any backflow between beats was blocked because it pushed the marble back on to the tube.

Although artificial valves did not arouse an immune response, the surrounding tissue grew over the implant as if to embrace it. This narrowed the diameter of the blood vessel, so the edge of the valve was honed into a blade so that when the marble rolled back, any overhanging tissue was chopped off. The valves also generated blood clots as blood cells got caught on the device, so the patient had to be dosed with anti-coagulants. Later, Teflon valves allowed a much smoother flow and they lasted for decades.

Other surgeons risked using heart valves from a dead donor. British surgeon Donald Ross made the first successful such transplant in 1962. He managed to reduce the failure rate from seventy-one per cent to fifteen per cent, but the shortage of donors and rejection problems remained.

Inventive medics decided artificial devices that were not prone to rejection were the way ahead. If the heart's own pacemaker becomes faulty, the body suffers a serious loss of oxygen. It had been known for centuries that the heart could be stimulated (or stopped) by electric shocks. In the 1930s Albert Hyman, a New York doctor, invented a large, hand-cranked device that generated a series of electric shocks that would restore the heartbeats to a regular rhythm. By the 1940s artificial pacemakers with two electrodes inserted beneath the skin did their best to sear the patient, who couldn't get away because he was tethered to an electric socket.

The invention of smaller electronics made miniaturisation possible. Two Swedish engineers, Rune Elmquist and Åke Senning, invented the first pacemaker that could be put *inside* the body. Now many thousands of them are implanted every year. Modern pacemakers are governed by microchips that adjust the pulses to the patient's activity and respond rapidly to an irregular heartbeat.

The idea of having an entirely artificial heart took root. As early as 1957 Willem Kolff, the inventor of the kidney dialysis machine, transplanted an artificial heart into animals, but its motor ran so hot that it cooked the recipients. In 1962 the National Institute of Health in the United States announced its intention to implant the first artificial heart into a human by 1970. Many inventors rose to the challenge and within a decade or so over a hundred patents for artificial hearts had been lodged. Kolff's team included a bioengineer called Robert Jarvik. He also experimented with animals and his Jarvik 7 model kept a dog alive for seven months. In 1982 the device replaced the failing heart of Barney Clark. He died sixteen weeks later from complications unrelated to the implant. Despite setbacks Jarvik pioneered artificial hearts ranging from early models that were the size of the demand valves on early aqualungs to tiny devices that could be easily implanted. The heart-pumps proved vital to keep patients with failing hearts alive until an appropriate heart donor was found.

Surgeons still had real hearts to mend. With heart surgery time was of the essence. Even a few extra minutes would help. In 1951 it occurred to Canadian surgeon Wilfred 'Bill' Bigelow that if he cooled patients down it would depress their metabolism, circulation and oxygen consumption and give the surgeon more time to operate. A year later Americans John Lewis and Walter Lillehei took up his idea. They operated on a five-year-old girl with holes in her heart chambers. She was wrapped in a blanket run through with tubes containing cold water and slowly her temperature fell to nine degrees below normal and her heartbeats slowed by almost a half. She fully recovered and there were no ill effects from the chilling. With the controlled hypothermia of patients modern heart surgery was born.

Bigelow had been trumped, but he had extracted a new

chemical from hibernating groundhogs that enabled them and human beings to tolerate much lower temperatures when under anaesthesia. He called the chemical Hiberin, but when he tried to patent it he got a surprise. It had already been patented as a plasticiser to make medical tubing more flexible and it was a strong alcohol. Bigelow had been plying his patients with booze.

Meanwhile scientists were tackling the problem of rejection. Sometimes implants had been inexplicably successful. In 1898 the young Winston Churchill, with no anaesthetic, sliced skin from his arm with a razor, to cover a comrade's wound. Apparently the graft 'took'. Modern researchers found that matching the tissue type of the donor with that of the recipient was as important as matching blood types; the closer the match, the better the chance of the transplant surviving. A laboratory in California has on file the tissue types of film stars awaiting the almost inevitable paternity suits.

Another approach was to tame the blood's white cells. At first doctors did this with X-rays and the patients were exposed to so much radiation that their bodies' defences were completely destroyed. A better way was to deploy anti-rejection drugs, but it was not until 1976 that a really powerful drug, cyclosporine, was isolated from a mould. Unfortunately, by suppressing the body's defences it made the patient vulnerable to any passing microbe.

The likelihood of rejection curbed the ambitions of heart surgeons, but by the 1960s they were competing to be the first to transplant a human heart. The prize went to an outsider in South Africa. Christiaan Barnard studied under Walter Lillehei and had visited several centres of heart surgery to learn more. He was ambitious, handsome and charming.

His interest in affairs of the heart was not confined to the operating theatre. Even on a brief visit to a London hospital he not only conferred with the surgeons, he also 'made friends' with a nurse.

In 1967 Barnard transplanted a heart from a young woman killed in a car crash into fifty-five-year-old Louis Washkansky, who didn't have long to live. Within a day the patient was alert and soon after he was out of bed, but eighteen days later he was dead. This apparent failure was lauded by the press and Barnard became the most famous surgeon in the world. He carried out five transplants a week, but within a few days the recipients were all dead. In the following year there was a veritable 'epidemic' of heart transplants world-wide. A hundred were carried out but the survival rate was so appalling that many surgeons gave up heart transplants.

They had rushed into surgery before there were adequate anti-rejection measures. It took another twenty years before heart surgeons could boast of a ninety per cent success rate. The secret was to reduce the anti-rejection drugs to the minimum suitable for the individual patient, just enough to keep infections at bay.

The number of heart transplants was determined by the availability of donors but the demand for organs far outstripped supply. An American surgeon became so exasperated waiting for a potential donor to die that he murdered him to harvest his heart. Others took diseased livers from cadavers, removed the tumours and transplanted the organ. Some patients were asked to sign a disclaimer agreeing to accept a sub-standard organ.

If healthy humans didn't volunteer to be donors, perhaps animals could be conscripted. In 1984 the heart of a baboon was transplanted into a desperately ill infant known only as 'Baby Fae'. Not surprisingly, a non-human organ provoked a big response from her immune system. She survived just three weeks and the surgeon received death threats from animal rights protesters. When asked how closely related baboons were to humans, he replied that his religious beliefs led him to suppose that humans were not related to animals at all. He never used animal organs again.

Other surgeons persevered and the animal of choice was the pig as its organs were about the same size as ours. They have been described as horizontal humans. Donald Longmore, a surgeon at London's National Heart Hospital, encountered unexpected obstacles. His patient was waiting for a heart transplant as soon as the donor pig arrived. Longmore intended to give him an extra heart to augment his diseased organ, a piggyback to keep him alive until a human donor was found. On arrival at the hospital the pig panicked and skedaddled down the road pursued by a posse of interns. A passer-by politely informed the pursuers that they were going the wrong way in a one-way street. The pig was captured, but a Jewish anaesthetist refused to take part. It was then revealed that the patient was also Jewish, so a rabbi was called in. The rabbi had a fit of the giggles when told of the dilemma. Eventually he composed himself and decided that as the pig might save the patient's life the implant was kosher. The operation went well until the surgeon injected calcium into the pig's heart to stimulate muscle tone. This was a routine procedure for humans but inexplicably it set like cement in the pig's heart.

Longmore never found out whether the pig's heart would have been rejected.

To overcome organ rejection researchers coated implants with alginate, a gel from brown seaweeds, which allowed nutrients and hormones to diffuse in and out, but made the organ 'invisible' to the patient's immune system. A more recent idea is to use a chemical secreted by parasitic worms to make them 'invisible' inside a body. A similar substance cloaks the developing foetus (which contains 'alien' genetic material from the father) to prevent rejection by the mother's immune system.

The most revolutionary idea is from Japan, where a surgeon named Hiromitsu Nakauchi plans to grow wholly human organs inside pigs. The pigs will be engineered to lack the desired organ, a kidney for example. Human stem cells would be injected into a pig embryo and the pig would grow normally with its own organs except for the kidney, which will be constructed of human cells. Nakauchi claims to have solved the rejection problem. In the United States the licensing body also allows human–animal chimeras so long as they do not breed.

An alternative is to grow entire organs in the laboratory using the patient's own cells, which the body would recognise as 'self'. It was easy to culture a thin sheet of cells, which was ideal for replacing a diseased bladder or a faulty heart valve. Growing a complex organ with all its blood vessels was a much more difficult task. But organs *can* be built using stem cells.

Our body contains a great variety of cells specialised for different tasks. A blood cell is very different from a muscle or nerve cell. But all of them are derived from stem cells which are undifferentiated. At the beginning of the twentieth century researchers discovered how to 'persuade' stem cells to change into almost any desired type of cell. They found a supply of stem cells in bone marrow and, more controversially, in foetal blood in the umbilical cord and placenta.

An organ was made by stripping a pig's organ of all its flesh using a mild detergent. This left a cell-less, sterile 'scaffold' of collagen on which the patient's stem cells were seeded. When cultured under the right conditions they divided and produced the appropriate specialist cells. The first person to grow an entire organ this way was Anthony Atala in the United States, who produced a bladder with its ureters and sphincter. The technique has also been used to produce a replacement windpipe, a womb and an intestine complete with muscles to squeeze the food along, and even the repair cells that continually replace the gut's lining. None of these was rejected. A cotton candy (candyfloss) machine was employed to make blood capillaries.

Recently 3D cellular printers adapted from ink-jet technology have built an inert scaffold for a human windpipe, but the slow growth and complexity of most organs suggest that it will be several decades before cultured organs end the reliance on human donors.

THE GENIE IN THE GENE

'They rightly do inherit heaven's graces'
- Shakespeare

Although humans have always known where babies come from, how they came to be as they are was a mystery. The unfortunate combination of a myopic philosopher and a microscope led to the sighting of a homunculus, a fully formed miniature being inside a man's sperm. Clearly women were merely incubators for a man-made child.

Mendel meddling

Although by the nineteenth century the woman's contribution was acknowledged, the inheritance of features was confusing and some characteristics skipped a generation and then reappeared. The first person to make sense of all this was a monk who liked gardening. The monastery at Brünn (now Brno in the Czech Republic) had a kitchen garden in which Gregor Mendel grew several different strains of peas. His aim was to cross them to make hybrids and to monitor the inheritance of obvious characteristics. He chose plants with contrasting features, such as purple or white flowers and smooth or wrinkled pods.

Mendel made sure they couldn't self-pollinate, then he transferred the pollen from the purple flowers to white ones and vice versa. When their peas ripened he sowed them. Surprisingly, all the next generation had purple flowers. Crossing these with one another resulted in offspring that were a mixture of purple and white flowers in a ratio of three to one. He got the same ratio for seven other characteristics.

Mendel rightly concluded that contrasting features were inherited separately and that purple was 'dominant' and therefore overrode white, which was 'recessive'. But in the third generation some plants received a double dose of the inheritable factor for white, so a quarter of the plants were white. He also assumed that the paired factors must separate in the reproductive cells (pollen), only to double again in the plant's ovule after pollination.

After ten years of experimenting Mendel published his results in 1865 and waited for the applause that never came. Darwin's *On the Origin of Species* had convulsed the world but Mendel's revelations created less excitement than a pea petal falling into a drain. Perhaps it was because he published his article in an obscure journal or because of its lists of numbers and ratios in an era of descriptive, not numerical, biology.

Disillusioned, Mendel concentrated on his religious duties and became the abbot. Sometimes he strolled through the garden with his pet fox, wistfully viewing the plants. He had discovered the mechanism of inheritance and no one cared. Years later the original manuscript of his publication was found in a wastepaper basket in the Brünn library. In 1900 Mendel's article was rediscovered and his achievements were celebrated. Surely Brünn's greatest son, a statue of him was erected in the square near the monastery despite local objections.

By 1902 chromosomes were shown to be the carriers of hereditary information and later genes were recognised as its repository even though they were far too small to be seen. Chromosomes were packed with proteins and nucleic acids but which one was the genetic material? It took American biologist Oswald Avery ten years to show that it was deoxy-ribonucleic acid – DNA. But how could a molecule made up of only four chemicals called bases hold the entire blueprint of life? The answer must lie in its structure and unravelling this enigma fell to James Watson, an American geneticist who came to Cambridge University, where he met Francis Crick. Watson was young and brash and considered most scientists to be narrow-minded, muddled and stupid. According to Watson, Crick was never in modest mode and always debated in a major key. He was also a brilliant theoretician too busy to finish his degree. Both the thinker and the go-getter were determined to break the DNA problem.

There were a few clues: chemical analysis of the four bases in DNA had shown that the amounts of adenine and thiamine were the same, while cytosine and guanine were present in equal amounts. Obviously they were paired. DNA formed large molecules, so these were probably crystals and could therefore be examined by X-ray diffraction, a technique that revealed what arrangements atoms make. It was like shining a light on a crystal chandelier and then trying to deduce the arrangement of the crystals from the pattern they threw on the ceiling.

In London Maurice Wilkins and Rosalind Franklin were

already looking at the DNA molecule. Watson and Crick's approach was to talk a lot and construct models to see how the bases might be held. The wily Watson snuck uninvited into a seminar in London at which Franklin was outlining her results. Watson's boss considered such behaviour was 'not playing the game', so he banned Watson and Crick from working on DNA. Naturally they took no notice.

The breakthrough came when, on a subsequent visit to London, Wilkins naively showed Watson Franklin's latest photo of the X-ray diffraction of DNA. Crick instantly realised that the pattern indicated a spiral and from the photo they could estimate the distance in which the helix made a full turn and also the number of cross-connections in each turn. This gave them the spacings their model had to match. The only configuration that fitted the data was a double helix like a ladder with the paired bases as the rungs. Crick's wife Odile drew a simple but elegant diagram of the double helix. To replicate life Crick and Watson envisaged the strands unwinding and then separating like the two sides of a zipper, each side taking its complement of the bases with it. Each half would then attract the appropriate bases to make two new double strands identical to the original one. It was the greatest scientific discovery of the twentieth century yet when their results were published in 1953 the public were completely distracted by the conquest of Everest and the Queen's coronation.

Watson wrote a racy account of the intrigues in the search for the double helix. Colleagues were enraged. Crick even accused Watson of violating their friendship. Watson dismissed Sir Lawrence Bragg, the head of the lab in Cambridge, as an out-dated curiosity. To give the book a stamp of approval

Watson asked the 'out-dated curiosity' to write a foreword
for it. Bragg was astonished at his gall, but his wife cajoled him
into complying. In the foreword he asks those mentioned in
the text to have a 'very forgiving spirit'. Rosalind Franklin,
whose photograph was the vital clue, was portrayed as a cold
fish. One of her colleagues said that though she was prickly
and difficult, he would walk through fire and water for her.
A poll in 2009 voted Franklin the second most inspirational
woman scientist after Marie Curie.

Publishers shied away from a book with such an appetite
for possible libel suits, but Weidenfeld & Nicolson bit the
bullet and it sold over a million copies. Its entertaining mix
of science, gossip and the quest for the double helix makes
Indiana Jones seem a very dull fellow. The pursuit of DNA
was frequently interrupted by the pursuit of local 'popsies'.
In a weak moment Watson even pondered what Rosalind
would look like without her glasses. He doesn't record what
the 'popsies' thought of the gangling and geeky-looking
American.

Once the structure of DNA was known, scientists sought a
way to locate specific genes and then to experiment with them.
It was not until 1973 that two American biochemists, Stanley
Cohen and Herbert Boyer, became the first genetic engineers
when they harnessed enzymes to 'cut' a strand of bacterial
DNA at a chosen place and then splice it together again. These
techniques enabled a normal gene to replace a faulty one or
to turn bacteria into millions of tiny factories producing useful
compounds such as an anti-malarial drug, insulin for diabetics
and a blood-clotting agent for haemophiliacs.

There is about 1 m (3 ft) of coiled DNA in every cell of
living things. Uncoiled, it is estimated the strand could stretch

from here to the sun and back again over six hundred times. But DNA has almost no bulk as it exists at the atomic level and initially molecular biology was hampered because scientists only had access to minute amounts of it.

In the United States Kary Mullis devised a system to increase the amount of DNA in a sample. He used enzymes to unzip and rezip double strands of DNA as required. This took place in repeated cycles of heating and cooling. Unfortunately, almost all enzymes are denatured by heat and the amplification process required temperatures as high as 95° C. This meant that after every cooling phase Mullis had to open up the apparatus to add fresh enzymes. As numerous cycles were required, the process was laborious, time-consuming and the enzymes were expensive.

One evening while Mullis was driving home it occurred to him that some bacteria live in hot springs and their enzymes must therefore be operating at high temperatures. In Yellowstone National Park there was a bacterium called *Thermus aquaticus* that had the enzymes he needed. His automated Polymerase Chain Reaction (PCR) could produce large amounts of DNA in a few hours. At every cycle the DNA doubled, so after thirty or so cycles a single molecule had multiplied several millionfold.

Mullis, who had earlier left science to become a baker, bought the enzyme for $35. Five years later he sold the rights to his invention for $300 million.

PCR amplification of DNA was imperative for scanning for genes that predisposed the body to disease. It even allowed scientists to examine the genetics of extinct animals by studying stuffed specimens in a museum. The best known use of PCR is genetic fingerprinting. It confirmed that the supposed skeletons of Tsar Nicholas II and his family were genuine and proved that Anna Anderson was *not* Anastasia. In 2013 Richard III, having never got his horse, had been hiding beneath a car park in Leicester for 528 years. His DNA was matched by a comparison with a living descendant in North America.

Leicester was also where genetic fingerprinting was invented. Alec Jeffreys, an enthusiastic and inspirational scientist at the local university, realised that although many of our genes are shared with all other humans, there are several regions where there is great variability. These regions display short sequences repeated over and over again. If several of the regions are examined the patterns are seen to be unique to an individual. The chance of two people (except identical twins) sharing all the same patterns is infinitesimal.

Now felons are detectable by a spot of blood, the sweat deposited by a fingerprint or the root of a single hair left at the crime scene. Such samples can also exonerate an innocent suspect. Genetic fingerprinting first solved a crime in 1987. There had been two separate murders in Narborough, just outside Leicester. Suspicion fell on Colin Pitchfork, a kitchen porter at a nearby hospital, who then confessed to one of the crimes, but not the other. The news of Jeffreys's invention had just broken and the suspect provided a blood sample for comparison with body fluids from the crime scene. The result was devastating as it proved that both murders were perpetrated by the same man, who was *not* Pitchfork.

The police decided to test every male in the district who was beyond puberty and not yet in his dotage, and they drew a blank. But one night some of the staff from the hospital kitchen were chatting in the pub. One fellow let slip that he had taken the blood test for Pitchfork. Indeed Pitchfork had excised his own photo from his passport and replaced it with the surrogate's so that it could be used for identification at the police station. The kitchen manager was so shocked she informed the police. Pitchfork was arrested, re-tested and charged with murder. He was given two life sentences.

In the UK every convicted felon now has his or her DNA added to a database. It identifies five hundred culprits a week, many the perpetrators of old crimes where the evidence could be revisited and tested. In the United States genetic fingerprinting revealed that one in eight prisoners on death row was innocent.

Jeffreys had struggled to get funding for his research, although he turned down the landlord of self-catering apartments who wanted him to identify who was wetting the beds. After the Narborough case there was worldwide demand for his invention, not just for criminal forensic work, but also for paternity disputes and plotting family histories. Now it is possible to trace one's genetic stock back to the original woman who gave rise to the entire human race.

It was not until the 1980s that the chemical code in DNA was read for the first time. In 2001 the draft mapping of the entire human genome was published. It identified over twenty thousand genes and the sequence of three billion base pairs. This made it much easier to locate specific genes such as those for cystic fibrosis, muscular dystrophy and haemophilia. The great hope is that we can harness genes to treat diseases that were considered to be incurable. Gene therapy has proved successful in the treatment of diseases of the immune system. In this endeavour viruses were the scientists' friend. Although we think of viruses as the harbingers of disease, they can be rendered harmless and used to carry vaccines and genes into the human body. The raison d'être of a virus is to invade a cell, insert its genes and take over the cell entirely. This behaviour has been used to deliver an enzyme into cells in the brain of Parkinson's sufferers to improve their motor skills. Viruses have also carried a gene that can stimulate nerve cells to produce a natural painkiller and it can deliver this to a specific location, unlike traditional painkillers that affect the entire body. Gene therapy is in its infancy but it is certain to revolutionise medicine in the near future.

THE LATEST THINGS

'Much I want which most would have'
– Sir Edward Dyer

Coming to our senses

Sight is our greatest gift and should not be abused. Isaac Newton was not aware of this. He stared at the sun to see if it damaged his eyes and had to retire to a dark room for days until his sight was fully restored. Not content with this, he pushed a bodkin (a long, blunt, large-eyed needle) into his eye socket and waggled it about at the back of the eyeball to probe what was there. He was lucky not to blind himself. Our sight deteriorates with age but defects can be corrected by spectacles. In Britain sixty-eight per cent of people wear them.

Emperor Nero watched the games at the Colosseum through a curved emerald to sharpen his view and to look cool. In 1285 a friar wrote: 'It is not twenty years since there was found the art of making eyeglasses which make a good vision . . . I have seen the man who first invented and created it.' A fresco painted by Tommaso da Modena in 1352 portrays an elderly cleric with spectacles perched on his nose. A contemporary, Sandro di Popozo, wrote: 'I am so debilitated with age that without the glasses known as spectacles, I would no longer be able to read or write.' Spectacles were a boon for scribes and scholars. With the advent of the printing press the demand for spectacles soared.

In 1727 Edward Scarlet, an English optician, was the first to secure glasses with side arms long enough to hook over the ears. Benjamin Franklin usually gets the credit for inventing bifocals. He commissioned a glazier to cut two sets of lenses in half and then cement them together so that he could clearly see both far and near. In fact bifocals had been invented by English optician Samuel Pierce nine years earlier.

For the active and the vain there are contact lenses. Leonardo da Vinci sketched contact lenses but it would be three hundred years before satisfactory lenses made them possible. The first practical contact lenses were invented by Dr Eugen Frick of Zurich, who designed them to correct distorted vision (astigmatism), but these and the hard plastic ones that followed were uncomfortable and couldn't be worn for long periods. It was not until 1960 that Czech chemist Otto Wichterle came up with lenses of soft gel plastic. Inexplicably the US Food and Drug Administration classified them as a drug, so they had to undergo extensive trials to be licensed. Where was the FDA when Newton was wielding his bodkin?

When I open a magazine an ad for sofas or hearing aids often drops out. I assume from this that we are a nation of half-baked couch potatoes with failing faculties. The ancient Greeks must have had the same problems, for they were often depicted lounging on a sofa. For hearing aids they used animal horns or shells of giant sea snails. The narrow end was sawn off and held to the ear to collect sound. In the sixteenth century an English physician named Thomas Willis claimed that background noise could improve hearing. He related the case of a woman who could hear speech only if a drum was beaten at the same time, so her husband detailed a servant to whack it whenever he wished to converse with her.

The first manufactured ear-trumpet came in 1624. Volta thought his electric batteries might help, so in 1790 he inserted metal rods into his ears and turned on the current. All he heard was the sound of 'thick soup boiling'. King John VI of Portugal had an acoustic throne with carved lions for arms. His courtiers had to kneel and speak into the lions' jaws. The sound travelled through the hollow arms to the king's earpieces. By the nineteenth century there were tubes aplenty snaking from an armchair to a trumpet in a vase in the houses of the well-to-do. One can imagine the farcical scene of a guest shouting into the flower pot and the host fiddling with his earpiece and swearing. Ear trumpets were concealed everywhere, in ladies' bonnets, tiaras and fans, or in men's top hats, walking sticks and beards.

Englishman Thomas Messenger's acoustic hat of 1900 had trumpet bells around the crown and females could sport Pauline Klaws's device of two years later which had two shells to fit round the ears attached to handles hooked into the hair. It was just the thing 'to be worn at a lecture or a concert', but I pity the people in the row behind them.

In America, the Dentophone was a fan-shaped 'sound collector' held in the teeth so that the vibrations from the speech of someone close by could be transmitted from the teeth to the ear bones in the recipient's skull. How the latter replied with a fan in his mouth is unclear. Payne's Magnetic Hearing Restorer was a pair of cylindrical magnets thrust into the ears 'to act upon the paramagnetic particles . . . of the blood and tissues'. There were several attempts at an artificial eardrum with a thread dangling outside to convey the sound waves to the drum.

Marconi's Otophone was the first electrically amplified 'portable' device, though it had to be lugged around in a large case. Miller and Hutchison in New York patented the Acousticon, which came with a large box filled with valves and batteries. At the coronation of Edward VII Queen Alexandra, who was hard of hearing, used the device to follow the ceremony. Truly portable hearing aids had to await the invention of miniaturised electronics.

Pens, paper and sticky stuff

For over a thousand years people wrote with a large quill pen: the feather of a goose or swan shaved to a point and split to provide the nibs. Although it looked elegant it had to be constantly re-shaved and the scribe had to tote an inkwell around with him.

Lewis Waterman from New York State was a travelling insurance salesman in the 1870s. When he offered his superior pen to a client to sign a contract the ink refused to run and shaking it splattered the document with ink blots. He rushed away to find a fresh contract but when he returned the client had gone. He found that the end of his pen screwed off, so he introduced ink into the hollow interior with an eye-dropper. Thereafter he would never dip the nib into an inkwell. He realised that air must enter the reservoir to prevent a vacuum developing and stopping the ink flow. It was the prototype of the capillary feed for his 'fountain pen', but it took years to develop one that didn't clog up or leak. Waterman was so hard up that he had to sell his patent and then pay for a licence to use his own invention. He seduced wealthy clients with engraved barrels, gold nibs and a clip so that it could be carried in the pocket. It was *the* pen to be seen with.

American John Loud lodged the first patent for a ballpoint pen in 1888, but it didn't flow reliably. László Bíró was a Hungarian sculptor, hypnotist and newspaper agent. He noted that newspaper ink dried quickly and cleanly. Unfortunately, newspaper ink was too thick to flow over the nib of a pen. It took years for Bíró and his brother György to devise a stainless steel, mobile ball that allowed the ink to flow when pressure was applied. Bíró sold the rights to the British pen manufacturer Miles Martin and the Royal Air Force took a large number because, unlike the fountain pen, the ink didn't leak out with the air-pressure changes. Sales did not take off until Bíró's pens were mass-produced by Parker in America, and France's Bic Crystal disposable pen company sold fourteen million a day. Bíró claimed that ballpoints wrote underwater and no matter what the angle. I tested both claims and was not convinced.

Richard Drew from Minnesota was a college dropout who made a living as a banjo player. By chance he met a painter who had just ruined a two-tone spray job on a car when the paint came off with the newspaper he'd used as a mask. Drew, now employed, solved the problem in 1925 by inventing masking tape and he didn't stop there. Cellophane was the new wrapping paper but how to seal it? Drew invented transparent sticky Scotch Tape (Sellotape in England and, confusingly, Durex in Australia). In its first year his company sold only $33 worth.

Art Fry worked for the same company as Drew. He was

in the church choir and helped out by putting slips of paper into the hymnals to mark the hymns to be sung. Not surprisingly, they often fell out. Their colleague Spencer Silver had experimented to find a super-strong adhesive when he stumbled on a very weak one. They developed a really thin layer of this wonderfully ineffective glue. The company was at a loss as how to market pads of not very sticky labels. To its surprise it was inundated with orders for Silver's Post-it notes.

A store owner in America asked researchers at the Drexel Institute of Technology to devise a way to automate the checkout and two graduate students, Bernard Silver and Norman Woodland, were intrigued. They invented a sort of visual Morse code of thick and thin lines. At first the lines were concentric circles as in a target but because there were only four of them the information they contained was limited.

They had the code but how could it be read? The code was in light-reflecting white lines on a black background which could be read by a photo-electric scanner. Their prototype was a big as a desk and took twenty minutes to scan a single item, by which time the 500-watt bulb had caused the wrappings to smoulder. It worked, provided the customers had unlimited patience and the proprietor was willing to risk setting fire to the shop.

RCA bought the rights and discarded the target in favour of lines printed in rows. With ten lines it was possible to code far more than just the name and price of the item. Woodland devised an identification code in which every grocery product had a unique number, and invented a laser to read it. In 1974 the very first item to be scanned in a store was a ten pack of Wrigley's Juicy Fruit chewing gum. It is now on display in the Smithsonian Institute's National Museum of American History.

John Davidson was not much of a gum chewer. He was a devout man who invented a device for monitoring appointments. The client wrote his name on an ivory wafer and it was slotted into a horizontal turntable. At the appointed moment the wafer fell and triggered an alarm to indicate your time was up. I wonder if Mr Davidson knew his invention was widely used in brothels?

For safekeeping

The frequency with which hordes of ancient coins and jewellery are dug up in farmers' fields suggests that burying precious objects was the safest option. Ships were a good bet as almost every wrecked galleon was said to be bulging with bullion and it was unlikely that a burglar could pilfer it from the ocean floor.

From the sixteenth century lockable, iron-bound 'Armada chests' and wrought-iron coffers relied on their weight for security. Thieves would be ruptured before they dragged the chest to the getaway cart.

Locks improved in 1818 when English locksmith Charles Chubb augmented the double-action tumbler lock with a 'detector' tumbler that jammed the lock if anyone tried to pick it. Chubb offered £100 and a pardon to a locksmith turned burglar if he could pick it. He couldn't. An American, Linus Yale Jnr., improved locks by giving each one a unique key. If by some fluke any unauthorised person managed to open a locked safe he might be surprised by Mr Elmer's toothed mantrap that lurched forward and clamped his hands until the police arrived.

Overlooking the sea on the promontory where I live in the

Isle of Man there is a memorial tower to Thomas Milner, who invented the first truly successful fireproof safe in 1840. It had a moisture-generating compound triggered by heat, in a watertight jacket lining the safe. When I was a lecturer at Glasgow University I invited a forensic scientist to give a talk to my students. One anecdote involved a burglar who used a wee bit too much explosive to open a safe. Its door was unhinged and the maker's brass plaque was blown across the room. It was likely that the thief was injured and a phone call to the local hospital confirmed this. A patient had a severely bruised chest and when viewed in a mirror it read:

> 'MILNER'S PATENT FIRE-RESISTING
> STRONG HOLDFAST SAFE.'

Danger was everywhere. The aptly named cut-throat razor had been manufactured in Sheffield since 1680. The slightest tremble of the hand caused blood to flow – the 'demon barber' Sweeney Todd was a great trembler, I understand.

King Camp Gillette's family lost everything in the great Chicago fire of 1871. He found little joy as a travelling salesman and longed to invent something to restore the family's fortunes. William Painter, who had invented the metal bottle cap, advised him to make something that was used and thrown away so that the customer had to come back for more. While shaving Gillette realised the open razor was far too long: only a small length was actually used for shaving. It occurred to him that a much smaller, disposable blade could do the job. He wrote to his wife: 'I've got it. Our future is made.'

He was a bit premature. Although he impressed sponsors with his idea, there were obstacles to overcome. He needed just a sliver of steel far thinner than any steel mill could roll.

So he turned to a clockmaker who had steel ribbons for springs. One of his sponsors was William Nickerson, who had invented the buttons for controlling elevators. He suggested that the razor would be safer if the blade were clamped against an outer guard so that just the edge of the blade was protruding. The final blades were as thin as the bristles they were cutting and a smidgeon of chromium in the steel enabled them to stay sharp for twenty shaves.

It took Gillette six years to bring his 'safety razor' to market in 1904. In the first year he sold only 168 blades, but a publicity campaign was launched featuring a baby shaving, with the caption 'Start early. Shave yourself.' Why would anyone use a cut-throat razor now there was a safety razor with cheap blades that never needed sharpening? In the second year he sold two thousand razors and over twelve million blades. Of course, there were a few diehards who stuck to the old ways. Oregon lumberjacks boasted that they shaved each other with their axes.

For millennia people had relied on pins and buttons to secure their clothes. In 1851 another American, Elias Howe, who would invent the washing machine, patented a device that was very like the modern zip fastener, but never marketed it. The next attempt was by a Chicago engineer, Whitcomb L. Judson, who invented the neatly named 'Clasp locker and unlocker for shoes' in 1893. As we all know, necessity is the mother of invention. Judson was portly and found it difficult to tie his shoelaces, so he designed a strip of metal with teeth that interlocked when drawn together. He had grasped the principle of the zip fastener but his heavy-duty device neither zipped nor fastened reliably and nothing is more embarrassing than a self-opening zip.

The numerous buttons on ladies' boots had to be fastened with a special button hook until a Swede named Gideon Sundback invented his hookless fastener in 1913. His Talon slide fastener was purchased for the US Army and a decade later the Goodrich Shoe Company adopted it for their galoshes. The company's director renamed it the Zipper and the name stuck. It was not until Schiaparelli in 1930 designed a gown with a zip from collar to hem that it gained respectability. Although a woman wearing such an easy-exit dress may not be all that respectable.

In 1948 Georges de Mestral, a Swiss aristocrat, was out shooting with his dog. The dog's fur became covered in burrs of burdock plants. De Mestral examined the burrs under a magnifying glass and was fascinated by the tiny hooks and the tenacity with which they clung to fur. An idea sprang into his brain and he enlisted a French weaver to produce a very unusual fabric. In 1956 he patented a new type of fastener, Velcro, from the French '*vel*ours' (velvet) and '*cro*chet', although there was no velvet in his marketed fasteners. The idea was simple: two strips of fabric, one with a myriad nylon hooks, and the other covered with tiny loops to snag the hooks. It could be fastened and released rapidly any number of times and it made a great ripping sound when undone. It's now everywhere from wallets to trouser flies – two items cherished by men.

A recipe for success

Isabella Beeton was one of the most famous women of the Victorian era. She was the eldest of twenty-one children but unlike most of her contemporaries she was highly educated, fluent in three languages and an accomplished pianist. At nineteen she married Samuel Beeton, a young journalist who made his name by publishing Harriet Beecher Stowe's *Uncle Tom's Cabin* in Britain. Their honeymoon was a tour of the leading European printers.

Isabella was so capable that her husband made her general editor of his *Englishwoman's Domestic Magazine*, in which she wrote a regular column on the latest Parisian fashions, including paper patterns which her husband invented, and recipes. A compilation of the recipes was published as *The Book of Household Management* in 1861.

As early as 1330 *The Forme of Cury* (not spicy curry but an archaic French word for 'cooking') included recipes. John Russell in his 1460 *Boke of Nurture* commented on food and confirmed what I have always believed: 'crabbe is a slutt to karve'. Beeton's innovations were that she listed the ingredients at the top of the page and specified the cost, cooking time and seasonality, and displayed the dishes in coloured illustrations. She offered plain food and occasional exotic items such as 'roasted wallaby' and 'parrot pie'. One recipe began: 'First catch your hare.'

All her books were aimed at the well-to-do middle-class wife with a servant or two. For example, the 'essentials' for a picnic were 'three corkscrews, four teapots, three-dozen quart bottles of ale (breast-feeding women should drink lots of beer), six bottles of claret, six of sherry, two of brandy and champagne *à discrétion*'. Beeton's books were not just about cooking: they were also about management. She saw the mistress of the house as a general who had to command and control her servants. The middle and upper classes could only operate because a bevy of servants took care of the everyday matters. They had to be invisible and were supplied with non-creaking shoes.

The young mistress of the house had no managerial training, so Beeton supplied it and gave her confidence. Her books tackled every aspect of family life, including child-rearing, first aid and legal matters. One book was all about gardening. Her advice was comprehensive, so readers could turn to her for everything from how to erase freckles to how to save someone struck by lightning. She invented the self-help manual that told the reader what to do and how to do it. And she coined the phrase 'a place for everything and everything in its place'.

Her first book was a blockbuster. It sold two million copies, but after that her publisher lost count. Her *History of the Origin, Properties and Uses of All Things Connected with Home Life and Comfort* was also a best-seller. When only twenty-eight-years old she died of puerperal fever (childbed fever) after the birth of her fourth child. *Mrs Beeton's Healthy Eating* was published in 1999, 134 years after her death. I hope it doesn't include the recipe in which she recommends boiling parsnips for one and a half hours.

Adventurous and stylish

Edwardian explorers required multi-purpose clothes. Tailors Holler and Ellis designed a cloak that also served as a hammock, a stretcher (poles included) and a float for crossing rivers. Lieutenant Halkett designed a stylish, waterproof cloak with an inflatable lining that could be transformed into a one-man dinghy within minutes. Its capacious pockets contained bellows and a paddle blade that would fit on to a walking stick. Halkett claimed to have tested his boat in the treacherous Bay of Biscay. An illustration shows him relaxing beneath an open umbrella to catch the wind. He also produced a collapsible two-man canoe. John Franklin took a Halkett boat on his Arctic expedition in 1845. He didn't return.

Mechanising the home

For decades the wealthy refused to modernise their homes. Even gas and electricity were considered vulgar, so many labour-saving devices were slow to gain popularity. The poor couldn't afford them and the wealthy had servants as their labour-saving devices.

The first device to thrive was the lawnmower. Landowners fell in love with meadows and lawns. The trouble was that the pesky grass kept on growing and aspired to be a jungle. The only tool that could effectively cut grass was the scythe, an instrument often carried by Death because unless in expert hands it is as likely to chop off a leg as trim a tussock.

A grand lady viewing her estate commented: 'A few sheep might make the fields look furnished.' Her husband enlisted sheep to mow the meadow but while the sheep chomps at one end it dumps at the other, which provides fine manure but is aesthetically suspect.

The first mechanical mower was invented in England by Thomas Plucknett in 1805. It was an industrial machine with carriage wheels geared to spin a large, circular blade held horizontally. It could decapitate unruly vegetation but not manicure a lawn. In 1830 an English engineer, Edwin Budding, filed a patent for a machine 'for cropping, or shearing the surface of lawns, grass plots and pleasure grounds', based on a device for shearing wool fabric. It had multiple blades arranged side by side on a horizontal shaft with wheels on either side and was promoted as giving 'amusing, useful and healthy exercise'. In practice it was a heavy brute to push and almost impossible to turn around at the end of a run. In 1899 John Burr of Massachusetts invented the first mower with curved blades that rotated when the machine was pushed. This prevented the clippings from clogging the gears, but it too was heavy to push up a slope.

Budding's mower had been manufactured by Ransome's Machining Company in Ipswich and in 1902 they produced the 'First and foremost petrol driven mower', which introduced a chain drive and a saddle for the driver. The company boasted of being 'Manufacturers of machinery to the King'. Mikhail Kalashnikov, though proud of his AK-47 rifle, wished he had produced something useful like a lawnmower.

Lawnmowers are the most dangerous domestic appliances. In the United States seventy-seven thousand mowing casualties attend an Emergency Room every year. Twelve per cent of them admit drinking beer while mowing. One toper from Milwaukee became the first man to detach his penis with a lawnmower. Don't ask me how.

A feature of houses was the incessant accumulation of dust and dirt. Brushes and feather dusters were ideal for disturbing settled dust and throwing it up into the air. Servants were given a carpet beater: a bamboo handle with a large shamrock-shaped paddle for whacking vast clouds of dust out of a carpet hung over a washing line. For larger carpets there was a carpet sweeper, a long-handled box with four wheels and revolving brushes. Carpet brushing was considered a housemaid's 'hardest torture of the week': she swept 'until every nerve is throbbing in fierce rebellion at the undue pressure to which it is subjected'.

The first vacuum cleaner was invented by London engineer Hubert Booth. Unimpressed by a demonstration of a blower to remove dust from railway carriage seats, he asked the demonstrator, why not *suck* the dust rather than blow it? The reply was, because it's impossible. That night Booth placed his white linen handkerchief on a chair cushion and sucked. It produced a grimy patch on the hanky, which convinced him that filtering out dust would work. He was used to building bridges and Ferris wheels, including the big wheel in Vienna that Orson Welles made famous in *The Third Man*. Not surprisingly, his vacuum cleaner was big. The power-driven pump came on a horse-drawn cart. It was parked outside the building while 800 ft (240 m) of flexible hose stretched into the rooms to suck dust from the carpets, tapestries and curtains. The motor was so noisy that neighbours complained and the first time the machine was used the horse bolted. However, it successfully cleaned the filthy carpets in Westminster Abbey in preparation for the coronation of Edward VII. The king was so impressed that he commissioned Booth to clean Buckingham Palace and Windsor Castle.

Several inventors produced more compact cleaners powered by bellows. A French engineer claimed that his machine 'seized dust particles from the air', but it seemed to expel the dust straight back into the air. Walter Griffith of Birmingham produced a domestic cleaner that needed two people, 'the maid and the maid's assistant', one to tread on the bellows the other to waggle the tube, which came with a choice of nozzles.

James Spangler, a janitor, blamed the dust thrown up from his carpet sweeper for his asthma. So in 1901 he built a vacuum cleaner out of an electric motor, a soap box, a broom handle and a pillow case. He gave one of his machines to his cousin, who was married to a saddle maker, William 'Boss' Hoover. As his main trade was declining with the rise of the automobile, Hoover bought Spangler's patent and put him on the payroll. The first electric 'Hoover' had revolving brushes to disturb the debris in the carpet and fan blades provided the suction to draw it into a collecting bag.

The competition was keen and catchy mottos were used to attract customers. The Ideal vacuum cleaner was 'Needed everywhere there was dirt', while the Aspirator plotted 'The doom of the duster, the brush and the broom'. But Hoover's 'It sweeps as it beats as it cleans' won the day, probably because Boss Hoover promoted door-to-door sales and home demonstrations. The salesmen were instructed to make sure that the house had electricity before they emptied the dust on to the floor to be sucked up. People called all vacuum cleaners a 'hoover' and the chore of 'hoovering' was born. Both Spangler and Hoover got rich, but only the salesman is remembered.

If there was anything worse than dirty carpets it was rotting food. The housewife or cook had to shop every day because, even in a cool pantry, food soon deteriorated. The ancients knew that food 'went off' on warm days and learned that drying, salting or smoking preserved meat and fish. In 1626 English polymath Francis Bacon tried stuffing chickens with snow to see if it preserved them. While collecting snow he caught a chill and died, thus ending his experimentation.

But how could you chill food in summer? The answer was

an ice pit. John Evelyn, the seventeenth-century English diarist, described how a deep pit was dug on the shaded side of a hill, lined with reeds and then filled with alternate layers of straw and winter ice 12 ft (3.7 m) high and thatched on top, to last through the summer. James I had brick-lined ice houses built at Greenwich and Hampton Court and by 1700 ice houses were commonly found in the grounds of large houses. In Scandinavia even relatively humble houses had them.

By the 1850s ice from Norway was taken to Britain in ships with insulated holds. Slabs of ice were delivered to houses and stored in a cold chest. It led to a craze for ice cream and sorbets.

Inventors had attempted to manufacture ice for many years. William Cullen, a chemist at the University of Glasgow, created freezing temperatures in 1775 by evaporating ether. As it changed state from liquid to gas it lowered the temperature of the air around it. But it was not until 1837, when James Harrison, a Scottish printer who had emigrated to Australia, was cleaning type and noticed that, as the ether he was using evaporated the metal type chilled, that its possibilities were realised. He built the first air-conditioner to cool his brewery in summer and to make ice. But his business failed and nothing became of his invention.

The first domestic refrigerator was the work of German engineer Karl von Linde in 1879. His steam-powered machine used a repeated cycle of compression to condense the refrigerant into a liquid and evaporation to remove heat. It was designed to chill beer and over twelve years he sold twelve thousand refrigerators. In 1923 Swedish engineers Carl Munters and Baltzar von Platen devised the Electrolux, which had an electric motor, and sold their patent to Kelvinator

Inc. in Chicago. Its noisy compressor was separate and kept in the basement. By the end of the 1930s only a quarter of British homes had a refrigerator as we were far less freezo-philic than Americans. But refrigerators had changed our shopping habits and our diet. As Edison's advert stated: 'It makes it safe to be hungry.'

The refrigerants were all toxic or corrosive substances such as sulphuric ether, sulphur dioxide and ammonia until American chemist Thomas Midgley Jnr. discovered the first of the freon gases in 1930. It evaporated instantly, was non-toxic and non-inflammable and so it went into every refrigerator. Unfortunately, when the fridge was decommissioned the freon escaped into the atmosphere and eroded the ozone layer that protects us from dangerous levels of ultraviolet radiation. It was not Midgley's first invention: he had already added tetraethyl lead to petrol as an anti-knock agent for smooth running. The consequence was that over seven decades billions of tons of toxic lead were expelled into the atmosphere. In 1980 eighty-eight per cent of American children under five had dangerous levels of lead in their blood. With the banning of lead in petrol the figure is now one per cent. Midgley contracted polio and could no longer walk, so he invented a hoist to get him in and out of bed, but he got tangled in it and accidentally strangled himself.

Canning was another means of preserving food. In 1810 London-based merchant and inventor Peter Durand produced a wrought-iron can in which food was sealed after being heated to destroy germs. It worked, but the cans were too heavy to be displayed on a shelf and their lead sealing was toxic and probably caused the death of Sir John Franklin's men on his expedition to chart the North-West Passage.

Surprisingly, the can-opener wasn't invented until thirty years after the can. The instructions were to use a hammer and chisel. Soldiers used a bayonet or even shot the can in frustration. The first can-opener was more likely to sever your wrist than open the can.

DIGITAL DOODAH

'I think there is a world market
for maybe five computers'
– attributed to Thomas J. Watson, chairman of IBM, in 1943

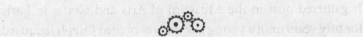

For five thousand years the human race has used devices to do sums. The abacus is still used by some aficionados who claim it is faster than a computer. Until the 1980s Japan was manufacturing two million of them every year.

The first mechanical adding machine was devised by nineteen-year-old French mathematician and philosopher Blaise Pascal. He built it in 1642 to help his father, a tax administrator. A handle turned a series of numbered wheels linked to gears that enabled numbers to be carried automatically and it could also subtract. Thirty years later a German mathematician, Gottfried Leibniz, who invented calculus independently of Newton, devised a calculator that could multiply and divide, but it was unreliable.

Punctured paper

Jacques de Vaucanson was a skilled maker of automata: novelty mechanical devices that moved under their own concealed power. The Frenchman's automated duck could quack, eat,

347

drink and even excrete. In 1741 he was asked to design an automated weaving loom. In those days a boy sat on top of the loom to lift the warp threads and not infrequently he made a costly mistake. Vaucanson installed a perforated strip of paper around a cylinder that instructed the loom to weave a particular pattern. The workers were not going to accept replacement by a roll of paper, so they sabotaged his machine. It gathered dust in the Museum of Arts and Crafts in Paris for fifty years until a young silk weaver named Joseph Jacquard was asked to rebuild it. He was intrigued by Vaucanson's device and replaced the paper roll with punched cards mounted on a belt that made it easy to select the required card. Each card had a pattern coded on it by the placement of punched holes. The loom could therefore weave complex patterns automatically. Jacquard's loom was also destroyed by workers fearing for their jobs, but the government compensated him and eventually his loom was widely adopted. Vaucanson's and Jacquard's looms were the first programmable manufacturing machines.

When the American west was still wild, outlaws posed as passengers on trains. En route they would don masks, rob the passengers and escape. To identify the culprits railroad companies recorded the features of every passenger boarding by punching holes in a card representing hair colour, facial hair and so on. If a robbery took place those who remained on the train were identified and those missing could be described to the police.

The 1880 US census took eight years to process, so the government's chief statistician was desperate to find a faster system. One of his staff, Herman Hollerith, thought the railroad's punched cards might be used to tabulate census information. All the data from the 1890 census was recorded on these cards and Hollerith built a machine that could read them. It had 'feelers' to detect the position of the holes and produce an electrical signal. The data was processed in half the time of the previous census even though the population had greatly increased. Hollerith's Tabulator was the first machine designed to process information and his system became the mainstay of office machines. His Tabulating Machine Company was the foundation of the computer giant IBM.

When I was a student punched cards and a knitting needle were used to store and retrieve large amounts of data. The police used punched cards to store all the information on the murders committed between 1975 and 1980 by the 'Yorkshire Ripper'. Punched cards were used as programmes for computers until 1960. They had numerous columns and each column had a pattern of holes representing one character. A hole meant 'yes' for that specific character.

We count in tens probably because we have ten fingers, but punched cards work on the principle of only two possible answers – yes or no. This is the binary system that your computer uses. Its programme codes are written in binary code composed of sequences of just two numbers, 0 and 1, represented electrically as 'off' and 'on'. The other two features of modern computers are that they can be programmed and they have a memory to retain data. The first machine with all these attributes was the brainchild of Charles Babbage, a brilliant mathematician who would invent the speedometer (odometer), the ophthalmoscope for examining eyes and the cow catcher for locomotives. While an undergraduate at Cambridge University he realised that logarithm and navigation tables calculated by hand were riddled with errors, so he decided to design a machine to do the job properly.

Mechanical mathematics

In 1823 while building his 'Difference Engine', which could handle numbers to twenty decimal places, Babbage decided to abandon it and began work on an 'Analytical Engine', a more powerful general-purpose machine. It would be one of the most complex and ambitious engineering challenges of the nineteenth century with a maze of fifty thousand cogs and wheels powered by a steam engine. It would also have a 'mechanical store' equivalent to a modern RAM memory. The complete machine would have been the size of a locomotive.

Babbage had the good fortune to meet Ada, Countess Lovelace, the daughter of Lord Byron the poet. Ada's mother Annabelle, whom Byron called his 'Princess of the parallelo-gram', encouraged her interest in astronomy and mathematics. Ada was fascinated by Babbage's vision and he called her the 'enchantress of numbers'. She wrote: 'The Analytical machine does not occupy the common ground with mere calculating machines . . . [it] weaves algebraical patterns just as the Jacquard loom weaves flowers and leaves. . . . It can do whatever we know how to order it to perform.'

Ada wrote all the instructions for the machine and punched a stack of cards to store the data. So she was the first computer programmer. She was only thirty-five when she died, but she has been immortalised by the United States Department of Defense, which named the Universal Programming Language Ada.

Babbage ran out of money and his sponsors withdrew because after all these years there was no completed machine, although he built part of one machine – a gleaming array of two thousand brass and gun-metal components. Instead of diplomatically explaining how important his machine was, he got angry and stated that the function of the Royal Society was 'to praise itself over wine and give each other medals'. Babbage was cantankerous and a professional complainer. He ranted about the street musicians who passed his house and proposed a bill in Parliament to have them banned. Neighbours booed him in the street and threw dead cats at him. Some even paid musicians to play outside his house.

Babbage returned to work, abandoning the Analytical Engine to build a better version of his earlier Difference Engine, but he died before it was finished. Fortunately, he left detailed blueprints of all his machines and in 1985 London's Science Museum began to build a 3-tonne (2.95 tons) part of his Analytical Engine. In 1991 it calculated its first equation to thirty-one decimal places. It has been called the most beautiful machine ever made. Experts believe that all Babbage's machines would have worked and it is accepted that he and Ada Byron designed the first proper computers.

Electronica

The future would not be mechanical machines but electronic computers. But the technology was slow and bulky, so the computers had to be big and so complex to operate that they were the exclusive domain of experts. In 1939 two Americans, John Atanasoff and Clifford Berry, used radio valves (vacuum tubes) to power the first fully electronic computer. They didn't patent their machine and willingly showed it to J. Presper Eckert and John Mauchly, researchers at the University of Pennsylvania, who then built their Electronic Numerical Integrator and Computer (ENIAC) with funding from the Department of Defense. It became a behemoth weighing over 30 tonnes (29.5 tons) and covering 15,000 sq ft (1,394 m²). Its innards, eighteen thousand valves, demanded electricity. When the machine was switched on it dimmed the lights of the entire neighbourhood. The operators worked in short bursts because it was always overheating to dangerous levels and a valve would blow about every seven minutes.

ENIAC was a thousand times faster than any previous computer, but every time it was reprogrammed thousands of switches had to be 'replugged' by hand. It also had the memory of an absent-minded gerbil, which was surprising as in 1940 Freddie Williams in England had discovered that cathode-ray tubes could provide random access memory (RAM).

Although ENIAC had been funded to track the trajectory of shells, it was not completed until 1946, by which time the urgency of missile monitoring had abated. Nonetheless, a replacement computer was commissioned and John von Neumann was brought in to supervise the project. He was a polymath and his calculations were vital in ensuring that the trigger mechanism would work in the plutonium bomb dropped on Nagasaki. He would be instrumental in promoting computing to the wider community.

Von Neumann described the new machine, the Electronic Discrete Automatic Computer, as a stored-program computer with a control unit, arithmetic unit, a 'considerable' memory, and a separate input and output. His detailed description of it became the blueprint for future computers. Ironically, it took so long to build that several computers with all its features were marketed before it was finished.

The most successful of the giant computers was Colossus. It was built in 1939 by Thomas Flower at the General Post Office laboratory and sent to the British military intelligence base at Bletchley Park. Its task was to break the Germans' codes. The war in the Atlantic was going badly because U-boats were decimating the convoys bringing essential supplies and *matériel* to Britain. Ships were being lost at a rate of almost sixty a month. Churchill later admitted that it was the one thing that really frightened him.

The German navy's Enigma code, invented by Arthur Scherbius, was so complex – it could configure fifteen million ways – that it was thought to be unbreakable. To crack it Bletchley used 200 computers called 'Bombes' that rapidly processed intercepted messages to and from U-boats, plus Colossus, which could process twenty-five thousand characters a second. It was so secret that its existence wasn't revealed for thirty-three years.

The other great advantage that Bletchley had was the greatest-ever code-breaker, Alan Turing. Eccentric and absent-minded, as befits a brilliant mathematician, he cycled to work wearing a gas mask for his hay fever and with his pyjamas beneath his jacket. He proposed to a woman at work and when she said yes he immediately told her he fancied men rather than women.

Turing invented the algorithm and Delilah, a scrambling device for telephone conversations, as well as designing both the Bombes and Colossus. With them he broke the Enigma code and the Tunny code with which Hitler communicated with his senior commanders. Without these successes the war could well have been lost.

Before the war, when a 'computer' was a person who computed, Turing envisioned the computer of the future. He defined a 'hypothetical universal machine', a new type of computer that was the exact one we have today.

By 1948 transistors, electrical on/off switches, could do all the tasks entrusted to radio valves. Geoffrey Dummer, a technician at the Royal Radar Establishment, had invented a flight simulator and a position indicator for planes in flight. He addressed a conference in 1952 and outlined electronic devices with no connecting wires but with transistors acting as conductors, amplifiers and rectifiers (to obtain direct current from AC supply). He envisaged an entire electronic circuit on a small wafer of silicon, but before it was built his funding was withdrawn. His boss could see no possible use for it as valves were more reliable. It was the worst decision his boss ever made.

Chips with everything

In 1958 Jack Kilby, an electrical engineer at Texas Instruments in Dallas, had realised that the era of elephantine computers was waning. He built an electric circuit using slips of silicon and capacitors (to store an electric charge), sliced from a power transistor and stuck on a sliver of germanium. It was the first integrated circuit, but his director was not impressed. There was no future for a circuit that cost ten times more than the existing technology. But Kilby persevered and within three years he had seventy components crowded on a single, small chip. His big break came when in 1961 the cold war threatened to get hotter. Ballistic missiles needed smaller, lighter and faster electronics and Kilby's tiny circuits were just the thing. A rocket's guidance system had over four thousand chips.

As these circuits work at the molecular level it wasn't difficult to miniaturise them. Since 1960 the number of transistors on a chip has doubled every two years. There are now millions of them on a microscopic chip, each component smaller than a virus. Without it there would be no laptop, cash card or mobile phone.

Kilby was not the only person making integrated circuits. Robert Norton Noyce worked at the laboratory of William Shockley, who had invented the transistor. Shockley treated his staff as if they were all spies trying to steal his ideas, so Noyce and seven colleagues walked out of 'Paranoid Place' and won backing from Fairchild Camera and Instrument Company to research semiconductors. Noyce had a risky strategy, to price his chips lower than the cost of all the components currently in use. This meant losing money but gaining customers. By the late 1960s mass production brought the manufacturing cost down and it was no longer an eight-man team: it employed ten thousand workers. Fairchild had the option to buy them out and it did.

Noyce and his partner Gordon Moore set up another company, Intelligent Electronics, abbreviated to Intel. In 1968 Ted Hoff, one of their technicians, invented the microprocessor, a single microchip that houses all the thinking parts and memory at the heart of your computer.

Caught in a web

In 1962 Paul Baran, a member of a US government think tank, was asked to find a way to guarantee that the military could control their missiles and bombers after suffering a nuclear attack. By 1969 the Advanced Research Projects Agency (ARPA) had Arpanet, a prototype internet that linked four universities.

The first message sent on the internet was 'Lo'. The sender meant to send 'Log in' but his machine crashed before he got to 'g'. It did that a lot in the early days. It was not until 1972 that Ray Tomlinson devised the first e-mail programme and sent the first heart-stopping message – 'qwertyuiop'. Bob Kahn of ARP developed a system that allowed any computer to send information to the internet. But finding the information was difficult until an Englishman, Tim Berners-Lee, came along.

Both his parents were mathematicians, so it was not surprising that their son constructed toy computers out of cardboard boxes and played complex mathematical games. While working at CERN (the European Organization for Nuclear Research) in Switzerland in the 1980s he found it difficult to communicate with other people in such a huge establishment. His solution was to create a programme using hypertext, the computer language of a viewing system that linked his computer to those of his colleagues.

In 1989 Berners-Lee released his hypertext to anyone who wanted it. If computers were to converse with each other they all had to speak the same language. He invented not only the simple computer language of the World Wide Web but also the addressing scheme that gives each web page a unique location and the first browser to enable us to get online. The first website went online in 1991 – a guide to Berners-Lee's creation, 'the Web', and how to navigate in it. Within five years there were forty million users; now there are billions. Berners-Lee didn't patent any of his inventions yet he changed the way we communicate and enjoyed the satisfaction of this and a knighthood from the Queen.

He hoped that the web would become a meeting place for inventors and scientists all over the world to pool their expertise and solve intractable problems. Recently 'crowdsourcing' has burgeoned. It is being used to sift through masses of raw data on, for example, astronomical bodies or the incidence of cancer cells on X-rays. It gives enthusiastic non-scientists a platform on which to contribute to scientific programmes. So if you are an amateur inventor, don't fret if you can't solve a problem yourself. There is someone out there who can help you.

A hundred and forty years ago Ada Byron foresaw that a computer 'might act upon other things beside number . . . Supposing . . . that the fundamental relations of pitched sounds in the science of harmony and of musical composition were susceptible of such expression and adaptations, the engine might compose, elaborate and scientific pieces of music.' How right she was. Music comes naturally to computers and in 2012 one called Iamus composed orchestral pieces in honour of Alan Turing. One composition, *Transits into an Abyss*, played by the London Symphony Orchestra, is available on CD.

COMING SOON

'I have seen the future and it works'
— **Lincoln Steffens**

Energy, where art thou?

I t is essential that sooner rather than later we cure our addiction to oil, coal and gas and adopt renewable resources. There are several options. Brazil's economy boomed on the back of growing corn for its sugar content, which was fermented to produce ethanol. Millions of cars in Brazil and the USA switched to run on ethanol. But to grow corn required land and water that could be better used for food crops. The crop required to fill the tank of an SUV (those gas-guzzling 'vans' on steroids) could feed someone in Africa for a year. The response to such criticism was to ferment non-edible plant waste such as corn straw, sawdust, paper and organic landfill. It is said that this process could replace all the oil the United States imports, but such claims are sometimes made to encourage investment.

The oil we tap from the geological strata today comes from tiny microalgae that died in shallow seas millions of years ago. The dry weight of some living microalgae is seventy per cent oil and some have been genetically engineered to produce diesel that, unlike conventional diesel, does not need to be refined. Other bacteria and yeasts have been engineered to put all their energy into ethanol production. They can be cultured in bulk and fed on waste water, so water purification is a by-product of biofuel production. Some biofuels give a better mileage than conventional diesel and with less greenhouse emissions.

The future of biofuel was given a fillip in 2011 when the US Secretary of the Navy announced that he would replace half the Navy's energy consumption with biofuels. This included fuel for helicopters and fast boats. In a trial a Hornet jet fighter using biofuel flew at almost twice the speed of sound.

There are several other possible alternative fuels that would run vehicles. During the Second World War ammonia was used. It doesn't release carbon, just water vapour and nitrogen, and it could take advantage of the current storage tanks and pipelines. Recently an ammonia-fuelled car cruised for over 111 miles (179 km) on just one tank. If the pioneering low-energy methods of producing ammonia are successful, it could be the fuel of the future. If not, there are now fuel cells that work on urine, so petrol stations could be replaced by drive-in urinals. Englishman James Highgate of Green Fuels shuns urine and petrol for cheese. His Lotus Exige sports car runs entirely on whey.

The US military is converting all of its home bases to solar energy and foot soldiers have small solar panels on their backpacks. Solar panels are light-harvesting silicon photo-voltaic cells that generate electricity. Until recently their efficiency was very low, though modern Grätzel cells are forty per cent efficient. A new type of cell has tiny antennae that resonate to produce an electric current when exposed to radiation. Unlike silicon cells, they respond to radiation on all sides. They are also sensitive to infrared radiation, which is almost half the spectrum. It can be collected even after the sun goes down as it is reflected from the earth at night. This should substantially increase the cells' efficiency.

Wind power is currently the front runner as an alternative to fossil fuels. It is a cliché to say that wind farms are blots on the landscape. Many years ago I saw the Altamont Pass wind farm in central California and I was entranced by a chorus of blades tracing graceful arabesques in the breeze. Public pressure, especially in Britain, has forced wind farms into the sea, making them far more expensive to build and difficult to maintain. Even so, they have now become competitive with energy from fossil fuels because their power generation has tripled thanks to more aerodynamic blades, better positioning to prevent mutual interference and turbines wound with superconducting wire instead of copper wire.

Not all the whirling blades are in the sky. Two rotors 53 ft (16.2 m) in diameter are submerged at the mouth of Strangford Lough in Northern Ireland. It is the world's first commercial-scale facility producing energy from the tide. Because water is much denser than air, an average tidal current produces ten times more energy via a turbine than an average wind turbine produces. The alternative to fixed turbines is devices harnessing the energy of waves. There are numerous weird and wonderful machines undergoing sea trials: buoys that bob up and down, a giant hinged 'oyster' that opens and shuts when repeatedly buffeted by the waves, an articulated 'train' 590 ft (180 m) long that writhes in the waves, and a water-filled 'snake', lurking just below the surface, that is squeezed by passing swells, creating a bulge of internal water that travels along the tube until it rushes through a turbine at the tail.

All these machines are at the mercy of ungovernable nature. What happens when the sun fails to shine, the wind doesn't blow or the sea is calm? There are ways of ensuring that the customer does not suffer any loss of power. At Port Dinorwic in Wales a mountain has been hollowed out. Inside is a set from the finale of a James Bond film, a cavern nine storeys high where enormous turbines spend most of their lives at ease. A languid technician leafs through the *TV Times* and, when *Coronation Street* ends, he flicks a switch so that water falls from the mountain lake above to the lake below and generates enough power to supply the millions of kettles that are about to be switched on. For the rest of the time water is gently pumped back up to await the end of *EastEnders*.

There is not always a handy mountain nearby, so a wind farm in Texas stores surplus energy in lead acid batteries the size of a double-decker bus. They store 153 megawatts to supply the grid instantly if needed. A solar array in Spain has 2,600 mirrors focusing on a central tower in which a solution of chemical salts is heated to 565° C. The heat is used to convert water into steam to turn turbines. Some of the solution is stored and, if needed, can supply electricity for up to fifteen hours.

Most solar arrays store energy as heat in huge, thickly insulated tanks, but phase-changing gel is the storage material of the future. It can absorb and release immense amounts of energy with very little change in its temperature. The Molecular Engineering Building at the University of Washington has the gel within its walls and ceiling spaces. The gel solidifies at night and melts in the heat of the day. This cools the interior by as much as ninety-eight per cent without air conditioning.

Before launching wind and solar energy into the grid, an energy supplier provided homes in Boulder, Colorado, with cheap smart meters for their refrigerators and air conditioners. The devices automatically switched off the appliances temporarily if there was a dip in supply.

Since the 1930s scientists have been trying to harness nuclear fusion, which could in theory provide limitless carbon-free energy. It is the process that powers the sun, which warms the earth even though it is 93 million miles (150 million km) away. Nuclear fusion is entirely different from the nuclear fission in present-day power stations. Fusion is fuelled by deuterium, which abounds in sea water, and lithium, which is used in laptop batteries. The major waste product is a small amount of helium.

At last, model nuclear fusion devices have been built. They prove that the process is possible, although they are far too small for energy generation, but in the future such devices could be used to transmute our stockpile of nuclear waste and reduce its radioactivity. Taylor Wilson was enrolled in a 'gifted students' programme when only fourteen and was given lab space at the University of Nevada. He is the youngest person to have built a fusion reactor.

To see if the process can work at the scale of a power station six countries and the European Union are funding the International Thermonuclear Experimental Reactor (ITER) being built in southern France. To confine the reaction, which is ten times hotter than the sun, it is surrounded by eighteen of the biggest magnets in the world. They have to be cooled to -269° C so that they become superconductors. Thus the coldest things on the planet will be within a few metres of the hottest thing on earth. By 2026 we will know if we can harness the energy of the stars.

Who's in charge?

Nuclear cars are not yet on the menu but major changes are imminent. Recently a British government minister went to prison because, when caught speeding, he lied that his wife was driving. But who would have been liable if no one had been driving?

The most dangerous people on the planet are not Al-Qaida or the Taliban: they are motorists. Every year around 1.2 million people worldwide are killed in road accidents caused by driver error. The car's most unreliable component is the driver. The solution to the problem is to let the car's computer take the wheel.

On April Fool's Day in 2012 Steve Mahan drove to his local Taco Bell eatery without touching any of the controls, which is just as well because he has only five per cent of his sight remaining. He was in one of Google's self-drive Toyotas – the first genuine automobiles. These have already travelled over 186,000 miles (300,000 km) without serious mishap. They navigate with sixty-four lasers continually scanning and a video camera to check the location against GPS positioning to confirm where the car is and to register road signs, traffic lights and pedestrian crossings.

The computer can memorise your regular journeys and do most of your driving and the driver takes over only in emergencies or if the system malfunctions. There is also a technology that enables cars to communicate to alert others of traffic jams or unexpected incidents. On motorways a convoy of cars linked by a type of Wi-Fi could proceed under the control of the lead car, to reduce fuel consumption.

Programming a computer to cope in busy traffic is no easy task. The Google car was instructed to give way at intersections, but this often led to it not getting a chance to proceed, so it had to be reprogrammed to wait a few minutes, then edge forward, signal and go.

Another problem is that a car's computer can be hacked. Two Spanish engineers have invented a device that bypasses the car's electronic control unit. It uses a $1 chip to break the encryption. It is quite possible that a hacker could take charge of a car remotely, perhaps with disastrous results. In 2013 Britain's High Court banned a researcher at the University of Birmingham from revealing the vulnerabilities of the ignition systems of Volkswagen cars, which would have been a boon for hackers.

Several states in the USA have passed laws to allow cars to steer, accelerate and brake unaided. Parliament is being pressured to bring in similar legislation in Britain. Inventors, engineers and the motor industry are keen to progress self-drive cars. Insurance companies are persuaded that self-drive cars will lead to fewer accidents because a driver constantly distracted by hands-free devices is accident-prone.

The initial reservation of passengers in these cars is down to the psychological effect of having no visible driver. I can envisage them having the look of suppressed terror that my wife exhibits when I take the wheel. The next thing may be pilotless aeroplanes. In 2001 an un-crewed plane with an artificial-intelligence navigation system flew from California to Australia. Don't be afraid: on long-haul flights the plane is on automatic pilot for most of the time so that the pilots can have a nap or conjugate with the stewardesses.

In the past futurists who predicted automatic chauffeurs imagined a manikin, not a box. By this time we should be serviced by humanoid butlers and android domestics to relieve us of our daily chores. The word 'robot' comes from the Czech *robota*, meaning 'forced labour'. Our robots would, of course, look like mechanical copies of us, except they would always do as they were told. Inventors have spent decades trying to get robots to do things that we can do with ease. Many mechanical maids have crushed china cups in their vice-like grip or poured tea on to the carpet. Countless androids on uncertain legs have tottered into chairs on an obstacle course that my two-year-old grandson would have skipped through. Why does a robot require legs? If it needs to move give it wheels.

Most recent robots are cute toys rather than useful machines. Their inventors strive to make them more human by covering their face with polymer skin and artificial muscles to animate their features. Strangely, the more lifelike the android, the more creepily sinister it becomes. The Fembot (female robot) is described as a slightly punk Goth with a look that would catch your eye across a crowded room. A romantic female android to replace the inflatable doll is said to be a sensitive, touching, companion. Curiously, the most charming and lovable robot is Wall-E, who resembles a tin safe on wheels.

The first useful robots were the spot welders that attended the production line for automobiles. Since then they have been programmed to talk, read signs and music and to sew. The most useful ones do jobs that are too dirty or dangerous for humans, such as defusing improvised mines. If it explodes, the robot's fingers are designed to fall off and can be readily reattached. Human fingers would also have been blown off, but would be more difficult to restore.

As robots mimic more and more of our characteristics inventors have tried to give them our intelligence. Can a machine think? In 1950 that question was addressed by Alan Turing, the catalyst for the quest for artificial intelligence. He stated that a machine could be said to be intelligent only if in conversation it cannot be distinguished from a human. No machine has so far passed this test. In the most recent attempt, in 2012 on the centenary of Turing's birth, Cleverbot was deemed to be human by fifty-nine per cent of the judges, but only sixty-three per cent of the panel thought that a human was human.

Machines can carry out some tasks better than we can. They can beat grand masters at chess and some supercomputers can perform complex calculations 100 million times faster than humans. But our brain requires only the energy of a dim light bulb, whereas the supercomputer needs 200,000 watts to do the same sum.

An intelligent robot should be able to cope in a busy environment. A recent robot called HERB honks as he approaches people. In a crowd he just stood in a corner honking. Researchers at IBM programmed a computer called Watson. They filled its memory with 200 million pages of text from newspapers and encyclopedias and entered him in the TV quiz *Jeopardy*. Pitted against two super-champions, Watson won, which was no mean feat because the questions were often subtle and oblique, containing riddles and puns. It can also learn as it amends its algorithms every time it gives a wrong answer, thereby increasing the chance of getting it right in the future. But Watson would fail Turing's test because in conversation it makes silly mistakes. It falls down on obvious things that everybody knows. Answering quiz questions is a feat of memory, not a sign of intellect.

Sony has produced a machine that can improvise music with live jazz musicians. The same laboratory has developed robots that have evolved their own language. A robot in front of a mirror adopts different poses and gives abstract names to each shape. Using that name, he then asks another robot to take up a shape. The second robot has no idea what pose has that name, so he chooses at random. If by chance he adopts the correct pose, the first robot nods his head and repeats the name. If it is the wrong pose, he shakes his head and demonstrates the correct one, repeating its name. The second robot adds this information to his database. Within a week a group of robots have evolved their own language to describe body positions which includes words for left and right.

Petti Haikonen, an inventor at the University of Illinois, has built an experimental cognitive robot (XCR) quite different from the rest. The cleverness of other robots is in their software, which would recognise the words 'pain' or 'blue', but would not *sense* them as we do. Haikonen's machine uses diodes and resistors so that the machine has real sensations sent to its 'brain'. For example, when hit it backs off, a response humans would exhibit. If it were holding something blue at the time, its blue diode responds and henceforth XCR associates blue with pain and avoids the colour.

There are now several robots that can make decisions for themselves and learn from their past experiences. Peter Dominey of the Robot Cognition laboratory at the French National Research Agency in Lyon has developed iCub, a robot that can track the mental state of others. This ability just 'emerged' without specific programming. This robot has what is known as a 'theory of mind' which in people is responsible for traits such as empathy. The 'Sally–Anne test' reveals this ability. A child is shown Sally putting her ball in a basket and then leaving. Next Anne arrives and moves the ball from the basket to a box and departs. When Sally reappears the child is asked where Sally will look for her ball. The iCub robot correctly answered: 'In the basket.' This skill is acquired by a child when about four years old. The inventors predict that iCub could lead to robots that can anticipate the needs of others.

There is a fear that robots will become super-brained human replicants indistinguishable from humans, as in *Blade Runner*. Androids have a long way to go if they are to develop human characteristics. We are amazingly imaginative and intuitive. Robots lack our warmth, humour, passion, empathy with others and those most human traits of all – common sense and irrationality.

Bionic man

As robots strive to become more human, people are becoming more bionic. The least intrusive but most robotesque is the 'exoskeleton' or 'muscle suit', a robot that can be worn. It has tiny motors at every joint to move the limbs, hands and feet so that a quadriplegic person can become mobile again.

Each mechanical finger has its own motor. It detects the shape of the object it is about to grasp and assumes the appropriate configuration.

The suits are activated by artificial muscles, rubber bladders worked by compressed air, or mechanical devices powered by hydrogen and alcohol. They can be far more powerful than human muscles, which would be an aid to anyone who has to lift heavy objects in their job. The suits have sensors so that the wearer feels sensations from the feet and hands.

The latest exoskeleton developed at the Santa Lucia Foundation Hospital in Rome has been trialled on a patient with a completely severed spinal cord. When he leans his upper body to one side a pressure sensor on the buttock moves the opposite leg. Leaning to the other side triggers the second leg to advance. In one test the patient wore a cap that measured electrical activity in numerous places on the skull, and he triggered some motion in the legs by thought alone. The objective is to develop the MindWalker suit.

Many human bodies are already fitted with artificial hearts, valves, bladders, tracheas and other organs. An eighty-three-year-old woman who lost her jaw from osteomyelitis was fitted with an artificial jawbone made by a 3D printer. Her muscles were attached to it and she was able to speak and eat normally again.

Prostheses for missing limbs are becoming more sophisticated. A mechanical ankle and foot can mimic the actions of the calf muscle and Achilles tendon to allow the wearer to walk, run and climb stairs with a natural gait. The Flex-foot, though nothing like the missing limbs, enabled Oscar Pistorius, a double amputee, to compete in the Olympics against able-bodied athletes. Initially he was banned as the bent blade prostheses were thought to give him an unfair advantage. His thighs provided the power, but in some prostheses the hand and arm can be activated by tiny electrical pulses generated by twitching the remaining muscle.

The devices can be controlled by thought alone. In 1872 Samuel Butler envisaged Utopia. One of his future-piercing prophecies was that machines would become 'nothing but extra-corporeal limbs'. His prediction became true in 2008 when two monkeys had electrodes implanted into the region of their brain that controls motor activity. The implant was connected to a computer operating a robotic arm. When a monkey fancied a marshmallow – typical monkey fare – a mechanical arm automatically picked up and delivered the treat. The monkey's brain had accepted that there was a third arm.

Four years later a woman who had been paralysed for fifteen years guided a flask of coffee to her lips with a mechanical arm controlled by her brain. Prior to this she had watched the 'arm' carrying out various tasks and she had to imagine that *she* was moving it. When she was doing this her brain signals were recorded by the computer and these were matched to the corresponding actions of the arm. Subsequently the arm responded to her thoughts.

Sensors inserted just 1 mm (0.04 in) into the brain use wireless signals to enable a patient to control by thought a wheelchair, a computer cursor, a TV remote control or a robotic arm. In 2012 an inventor manufactured two semi-autonomous arms strapped on to a harness. They were designed to help the wearer when an extra hand or two was required. The arms protruded forward at waist level. Goodness knows what the wearer's girlfriend thought the first time handy Andy approached. The hands were programmed to use power tools, but anyone who has seen *The Texas Chainsaw Massacre* knows what happens when power tools get into the wrong hands.

SCIENCE FICTIONS?

'I dipped into the future . . . saw all the wonder that could be'
– Alfred, Lord Tennyson

'Put no trust in the future'
– Henry Wadsworth Longfellow

A mong the most inventive people are writers of fiction who envisage the future. Cyrano de Bergerac had men rocketing to the moon in 1657. In 1862 another Frenchman, Jules Verne, predicted that the world in 1960 would have skyscrapers, an internet and fax machines. His book was turned down by publishers as too fantastic. Orwell's *1984*, written in 1949, had indispensable inventions such as brainwashing, constant camera surveillance and a national lottery. In 1828 Jane Webb had an Egyptian mummy being brought back to life. Nonsense, of course . . . Or is it?

Raising the dead

You might think that the dead are obviously so. They do not breathe, their heart has fallen silent and their brain has ceased to spark. But there has always been a nagging doubt that we might be buried prematurely and would awaken in the dark confines of a coffin six feet under. To avoid such a mishap bodies were kept until the aroma left no doubt as to their condition. Even so, there were numerous stories of shroud-bursting cadavers. A missionary lying in an open coffin during his funeral service suddenly joined in with the choir. Apparently it was his favourite hymn. Dozens of corpses are said to have sat up, to the surprise, and possible dismay, of their relatives. Sadly, sitting up may not alert the congregation when the lid is screwed down.

To reduce such incidents the diagnosis of death had to be more rigorous. The recommended tests included blowing pepper or jamming a sharp pencil up the deceased's nose, whipping with stinging nettles or the surprise of a hot poker up the rectum. That was sure to get a response even if they *were* dead.

For the unconvinced there were several models of 'security coffins'. George Bateson's Life Revival Device of 1852 involved a cord tied to the hand of the deceased. It passed through a hole in the coffin lid and up to a bell on a scaffold. Should the corpse revive he would pull the cord and ring the bell in the hope that someone with a spade was within earshot. If the grave digger was hard of hearing, it was just tough luck and he had missed an opportunity to prise open the lid and say: 'You rang, sir?'

Bateson was awarded the Order of the British Empire for 'services to the dead'. His faith in what was commonly known as 'Bateson's belfry' can be judged by his insistence that he must be cremated, not buried. An American version was the 'deliverance coffin', which fitted the corpse with a bridle so that the slightest movement released a spring in a pipe which in turn thrust what resembled a floor mop into the air and also erected a flag. Unfortunately, the gases released by decomposition caused sufficient movement to hoist the flag and mop and the grave digger would waste his time.

Security coffins were still available in the 1970s. The upmarket model was equipped with a wireless to call for help and a hamper so that the hungry corpse could have a picnic while waiting to be exhumed. For those who cannot afford such luxuries I advise leaving the lid off the coffin until the very last moment, make sure the hymns are loud and, if all else fails, fall back on a warm poker.

In the late-Victorian era, when houses were lit by flickering gas lamps and few people could swim, asphyxiation was a common cause of death. The Royal Humane Society attempted to revive the drowned. Along the banks of the River Thames they hung survival equipment. It consisted of a pipe attached to bellows that would propel tobacco smoke up the patient's bottom, perhaps in lieu of a final cigarette.

The brilliant eighteenth-century surgeon John Hunter correctly believed that the dead were just undergoing a suspension of physiological activity. To revive them he advised pumping air into the lungs and restarting the heart with electrical shocks. Two centuries would pass before defibrillators became widely available.

In the late nineteenth century George Poe, a cousin of Edgar Allan Poe, devised life-saving equipment to revivify the dead. It pumped oxygen into the body and pushed air out. In public demonstrations he asphyxiated dogs and rabbits and when they were dead he brought them back to life. One poor rabbit was asphyxiated and revived eleven times. A woman in the audience offered to be the first human to be retrieved from the dark veil, but Poe wisely demurred.

Today our ability to detect different aspects of human physiology has blurred the distinction between life and death. Typically, a patient arriving in hospital with heart failure is given CPR (manual chest compressions) for around thirty minutes to keep the body aerated – mouth-to-mouth resuscitation is now out of fashion. A defibrillator is used to restart the heart. If by this time the patient is not breathing without help it is assumed that the brain cells starved of oxygen are beginning to die irreversibly and the patient is usually pronounced dead. CPR is successful in only five per cent or so of patients.

However, some surgeons have reversed death in patients who have been dead five or more hours because they had had the right treatment. CPR is helpful before reaching the hospital but as soon as possible the patient should be linked up to a machine that takes over their breathing and an ECMO (extracorporeal membrane oxygenation machine), which siphons blood from the patient, removes carbon dioxide and sends oxygenated blood back. It is critical to keep the oxygen level above forty-five per cent of normal, or the heart won't restart. The third requirement is to cool the body, and especially the brain, by injecting cold water into the nostrils for example. The brain is chilled to reduce its activity and its need for oxygen, so it can survive without the brain cells being impaired. This gives the surgeon time to diagnose the patient's condition and perhaps start remedial surgery. It is estimated that these procedures could save forty thousand lives a year in the United States and ten thousand in Britain.

We don't yet know how long the dead retain their psyche, the mind and soul that constitute a person, but it may be longer than we thought. However, the benefits of cooling the human have been taken to absurd extremes. Perhaps American physician Robert Ellinger was inspired by the way frozen peas lasted in the freezer. He thought that the dead could do the same. He treated the recently deceased by replacing their blood with oxygen-enriched fluids or even anti-freeze and sealing them in a vat of liquid nitrogen at -196° C. These 'patients' had to be very patient indeed to await being resuscitated in the distant future when the diseases from which they died could be cured. Perhaps Ellinger had overlooked that even frozen peas have a 'best before' warning.

Chilly confinement in the hoarfrost hotel costs a cool $150,000 plus an annual maintenance fee. A cheapskate can opt for just conserving his head for $90,000, but that may be a false economy. The rich think they can buy everything, even immortality. One customer was so confident that he willed all his wealth to himself.

Ellinger died in 2011 and now rests in a sealed can next to his mother, two wives and their dog – a cosy family gathering of petrified popsicles.

What will remain in the far distant future when resurrection time arrives? Human bodies are almost entirely composed of water. When frozen, ice crystals shatter cell membranes and disrupt all the vital organelles inside the cell. Only super-fast cooling can moderate crystal growth. What will finally emerge from the canisters and the customer is thawed out? My bet is a very bad smell.

Never say die

Many of us feel we've been given short measure. Adam, Noah and Methuselah lasted up to 900 years, although because they all fell short of a millennium, they probably felt cheated too. In 2006 a Galápagos tortoise who knew Clive of India finally died. But no one wants to live that long if they have to look like a tortoise.

There are reputedly twenty thousand centenarians in Japan (about a tenth of all those in the world), with the highest concentration on the island of Okinawa. It isn't something special in their genes, for locals who emigrate to the west, like those who left Shangri-La, lose their longevity. Anyone in search of the fountain of youth should emigrate to Okinawa. If already decrepit, they might try rural China, where centenarians are reported to be enjoying restored hair colour and growing new sets of teeth. A bald man is now completely re-thatched and one game old girl has restarted her menstrual cycle. In 2007 Pan Xiting (aged 106) married a young lass of eighty-one. A local official confided that she means 'more than a companion' to Mr Xiting.

For a long life, make sure you are a woman and the first-born of a mother less than twenty-five years old; it doubles your chances of living to a hundred. On average, happily married couples live far longer than singletons; university graduates, if they studied medicine rather than arts, fare better in the longevity stakes than non-graduates; those with a Master's degree live longer still and PhDs longest of all. If you manage to win an Oscar you are likely to live three years longer than the losing nominees.

Mankind has always longed for the fountain of youth and some thought they had found the route to rejuvenation. Women knew that a man is only as sensible as his testicles allow, but Charles-Édouard Brown-Séquard, an ageing Mauritian physiologist, decided that a man was only as old as his testicles. While others gave their glands an electric fillip with electrodes, he injected himself with extracts of ram's testes. He proudly announced to his students: 'Gentlemen, last night I was able to pay a visit to Madame Brown-Séquard.' Perhaps he loved her to excess, for a few weeks later he died.

Injections of goat or monkey gland and even testicle implants became fashionable rejuvenators. Sigmund Freud was a beneficiary. Ironically, eunuchs lived seventeen years longer than their testicled contemporaries. Some of the injectees' wives conceived and the doctors were relieved to report that the babies had 'no simian symptoms' – at least no more than the father displayed.

Today the race is on to extend our lifespan. We have known for over seventy years that mice live over fifty per cent longer on a calorie-restricted diet. Rhesus monkeys kept on thirty per cent less calories (but the same nutritional value as their 'normal' diet) for eighteen years now have 'younger' cardio-vascular, metabolic and immune systems than others of the same age. These are the very systems that often let us down in old age. In 2001 in America human volunteers began a self-experiment: a lifetime on two-thirds rations. They will be regularly monitored for signs of age-related degeneration. Assuming they stick to their diet for the rest of their days and don't succumb to the occasional Mars bar, time will tell whether they live significantly longer than the rest of us. Even if it works, who would savour a life feeling hungry all the time and having insufficient energy to chase a woman? A side effect of this regime is a lack of libido, so even catching her would be an anticlimax. The most recent idea is to eat conservatively, shun red meat and fast once a week.

The mechanisms that underlie the process of ageing are

hidden deep within our tissues and are only now being unveiled. The clocks that tick away our youth are chromosomes. They have caps of DNA on either end to protect the tips, like the plastic collars on the ends of shoelaces. Over time the caps erode and this is somehow linked to the ageing of the body. People over sixty with shorter caps were almost nine times more likely to die from an infectious disease. A large study of men found that those with shorter caps were more prone to heart attacks. Smoking, drinking and obesity accelerate the shortening and when the caps get very short it can lead to cell death.

The loss of caps is not inevitable: telomerase, an enzyme that encourages cell division, can restore them with new DNA. Trials with humans given nutritional supplements that stimulate telomerase produced a substantial decrease in patients with small caps. With telomerase the lifespan of human cells in a laboratory culture increased fivefold and some cultures continued to divide for over three hundred generations beyond their normal life expectancy. In the future telomerase may become a major rejuvenator to restore our ageing bodies, but as it stimulates cell division so its ability to create tumours must be curbed.

The role of genes in ageing is receiving considerable research. It is thought that the genetic contribution to lifespan is ten to thirty-five per cent. Disabling a gene called daf-2 allowed mice, worms and flies to live more than double their typical lifespan. Daf-2 causes a cascade of other genes to be switched on or off, resulting in better protection and tissue repair, thus allowing the animals to live longer. Many centenarians have a mutation of daf-2. There is also evidence of a longevity gene in fruit flies that prevents cells from self-destructing under stress and improves communication between nerve cells. No doubt a similar gene is being sought in humans.

It is now possible to disable the genes for many of the common diseases that kill us before we reach a great age. Stem cells can repair and replace ageing organs and tissues. We are also on the brink of extracting cells from the patient, genetically engineering them to correct a serious defect before returning them to the patient's body. We are definitely on the mend.

There have been many false dawns on the road to retarding ageing and extending human lifespan. When free radicals were all the rage high doses of resveratrol, a constituent of grapes and red wine, were found to prolong the life of fish, inspiring many people to turn to the bottle. Unfortunately, the dosage given to the fish equated to taking seventy-two bottles of red wine a day. Now that really is drinking like a fish.

The control centre for ageing is the hypothalamus in the brain. Experiments on mice introduced into the hypothalamus a protein complex that regulates the immune system. This treatment extended the lifespan of mice and the animals remained fit and mentally acute longer. The protein complex reduced the production of a gonadotropin-releasing hormone that stimulated the production of new neurons in the brain. Ageing was slowed down even in mice genetically engineered to age rapidly. This suggests that the hypothalamus is the gateway for age-controlling drugs and the best bet for increasing a person's longevity. But the hypothalamus controls most of our body's systems, so we don't yet know what might happen if we meddle with it.

Over the past fifty years there has been a fivefold increase in Britain in the number of people aged over eighty-five. The average life expectancy of a girl born in 2007 is calculated to be eighty-one and just over seventy-six and a half for a boy. This is based on the fact that life expectancy has risen by 2.2 years every decade over the past hundred years. But things may be about to change. Our sedentary, flab-enhanced youth are likely to live shorter lives than their parents. Cirrhosis of the liver lies in wait for many of them. The elixir of youth is definitely not to be found in a cider or vodka bottle.

Living longer *is* possible. A French lady named Jeanne Calment made it to beyond her one hundred and twenty-second birthday. Most scientists in the field believe that they can fashion a fit and healthy life extending to Madame Calment's age or even more. But immortality is not just around the corner and even if it were possible it is not desirable. My descendants are my immortality. Death is just the price we pay for the privilege of having lived.

George Santayana said: 'Having been born is a bad augury for immortality' and who could disagree with the summation of a contestant in the Miss USA beauty contest: 'I would not live forever because we should not live forever, because if we were supposed to live forever, then we *would* live forever, but we cannot live forever, which is why I would not live forever.'

Creating life

For the ancients the origin of life was well known. Fleas came from filth, toads from mud, and mice were generated by putting damp rags in a pot with a scattering of wheat germs.

In the seventeenth century a Florentine scientist proved that these and many similar beliefs were myths, but the Bible confirmed that all it took to create a human being was a divine magician and 'dust from the ground'. The belief that life could spring from non-living matter survived into the Victorian era.

It is, of course, obvious that life begets life, but the very first life form had no parent: it arose in a lifeless world. In 1828 urea, a compound found in urine, was synthesised from chemicals. It was the first evidence that the transformation of inorganic to organic material was possible.

Charles Darwin suggested that life might have arisen in a warm pond with compounds such as ammonia and phosphoric salts plus light, heat and electricity in the form of lightning. In 1953 American physical chemist Harold C. Urey was lecturing to students on the birth of life from chemicals and challenged them to design a laboratory experiment to test if it were possible. Stanley Miller rose to the bait. His bubbling flask of chemicals intermittently zapped by an electric spark was eerily reminiscent of horror films in which a mad scientist strives to create life. The flask contained what was thought to be the atmosphere of our lifeless planet billions of years ago – a cocktail of water vapour, ammonia, hydrogen and methane. To Miller's delight this inorganic brew produced amino acids, the building blocks of proteins. Further experiments delivered nineteen of the twenty protein-producing amino acids that occur in living organisms.

Sixteen years later a meteorite fell in an Australian desert. A large fragment determined to be 4.5 billion years old contained seventy-nine amino acids and also fatty acids that can form cell membranes. Furthermore, the amino acids had been produced by exactly the same sequence of chemical reactions as in Miller's flask.

We now know that life originated much earlier than Miller imagined and the atmosphere at the time was different. There was probably less hydrogen, but more nitrogen, carbon dioxide and water. If Miller's experiment had used this atmosphere, it would not have generated amino acids. Scientists have irradiated the revised ancient atmosphere with ultraviolet light and it produced a substance that is involved in the chemical cycle that all living organisms employ to create energy. Biochemists have convincing scenarios of how the early cells and photosynthesis developed, but what ignited that very first spark of life still eludes them.

Creating life, however, may soon cease to be the monopoly of Victor Frankenstein. A new science called synthetic biology empowers scientists to manipulate genes and redesign life to whatever is desired. For example, a female goat called Freckles now has an implanted silk-making gene from a spider. The strands of silk are expressed in its milk and can be spooled off, leaving the milk to nourish Freckle's offspring. The goat is a means to mass-produce a potentially useful material that is far stronger than Kevlar. But it's more freaky than freckly.

In the past most inventions have been mechanical or electronic. In the future the big and most daring innovations will come from synthetic biology. In 2010 Craig Venter, an American biologist, announced that he had produced Synthia, a synthetic life form. He had stitched together bits of chemically synthesised DNA and inserted them into an empty bacterial cell. This synthetic cell was then booted up and as ordered by its set of instructions it produced billions of exact copies of itself. He claimed to have created a new species.

The press berated Venter for 'playing God'. But the verdict of some fellow scientists was that although it was a remarkable feat, he had not created life, merely rebooted it. Others, more cynical, believed that his patent was an attempt to monopolise all the techniques of genetic manipulation. We have survived artificial breasts, synthetic flavouring and synthetic music. Synthetic life may be just around the corner.

Laboratories around the world are synthesising many of the processes that were the prerogative of living organisms. DNA and RNA are the means by which living things send genetic material to their descendants, but there are now six synthetic molecules that can pass on genes to another artificial molecule. There is also a device that can start or halt the flow of an enzyme that synthesises RNA, much as a transistor controls the flow of electricity. This opens the door to living gadgets. As we all know, orgasms can be faked but even sex can be synthesised. For mice both eggs and sperm can be grown from scratch. Researchers are aiming to adapt their methods for humans. It should be possible for one person to supply the stem cells to make both sperm and eggs, bypassing sex altogether. No doubt there will be some stick in the muds who prefer to make babies in the old-fashioned way.

Surveys in both the UK and USA have revealed, surprisingly, that the majority of the public are not concerned about the possibilities of synthetic biology, and indeed welcome the medical advances that might ensue. About a third of those consulted were more cautious and even suggested a moratorium until the consequences become clear. Creating artificial life may not be the best way to reassure people.

You can now buy BioBricks from an online catalogue. They

are bits of DNA designed as tools to do specific jobs. Like LEGO, they can be assembled on to a machine – you just mix the bricks and add glue. There is concern that such articles are freely available. Goodness knows what the recipient might build and goodness may have nothing to do with it.

The future holds unimagined wonders, but fasten your seat belt – it's going to be a bumpy ride.

REFERENCES

INVENTIONS BEFORE THERE WERE INVENTORS

Barras, C., 'Our Asian origins' in *New Scientist*, 11 May 2013, 41–43

Bowles, S., 'History lesson from the first farmers' in *New Scientist*, 30 July 2011, 26–27

Brahic, C., 'I cook therefore I am . . .' in *New Scientist*, 17 July 2010, 12

Brown, P.L., *Megaliths, Myths and Man*, Book Club Associates, London, 1977

Callaway, E., 'Modern humans' Neanderthal origins' in *New Scientist*, 15 May 2010, 8

Channel 4 TV (UK), Pearson, M.P., *Secrets of the Stonehenge Skeletons*, 10 March 2013

Gore, R., 'The dawn of humans: Neandertals' in *National Geographic*, January 1996, 2–35

Gotaas, T., *Running: A Global History*, Reaktion, London, 2010

Green, R.E., *et al.*, 'A draft sequence of the Neandertal genome' in *Science*, 7 May 2010, 710–722

Hamzelou, J., 'First Europeans did not rely on fire' in *New Scientist*, 19 March 2011, 16

Hart-Davies, A., *What the Past Did for Us*, BBC Books, 2004

Hawkins, G.S., *Stonehenge Decoded*, Fontana, London, 1970

Hetch, J., 'Evolution made us marathon runners' in *New Scientist*, 20 November 2004, 15

References

Holmes, B., 'Manna or millstone' in *New Scientist*, 18 September 2004, 29–31

James, P., and N. Thorpe, *Ancient Inventions*, Michael O'Mara Books Ltd, London, 1995

Leakey, R.E., *The Making of Mankind*, Book Club Associates, London, 1981

MacKie, E., *The Megalith Builders*, Book Club Associates, London, 1977

McKie, R., 'The vagina monoliths: Stonehenge was an ancient sex symbol' in *The Observer*, 6 July 2003, 7

McKie, R., 'As rival theories tumble, mystery of Stonehenge keeps us guessing' in *The Observer*, 17 March 2013, 20–21

McKie, R., 'Big strong, great sight – but it wasn't enough to save the Neanderthals' in *The Observer*, 2 June 2013, 28

McNeil, I. (ed.), *An Encyclopedia of the History of Technology*, Routledge, Abingdon, 1990

Marshall, M., 'Bear DNA could give away cave art's age' in *New Scientist*, 23 April 2011, 10

Marshall, M., 'Out of Africa, into bed with the locals' in *New Scientist*, 18 June 2011, 11

Mithen, S., *After the Ice*, Phoenix, London, 2003

Newham, C.A., 'Stonehenge: a Neolithic observatory' in *Nature*, 211, 1966, 456–468

Renfrew, C., *Before Civilization*, Penguin Books, Harmondsworth, 1976

Shreeve, J., 'Limbs ancient and modern' in *The Independent on Sunday*, 14 April 1996, 48

Stokstad, M., *Art History*, vol. 1, Prentice Hall and H.N. Abrams Inc., New York, 1995

Thom, A., *Megalithic Lunar Observatories*, Oxford University Press, 1971

Thom, A., and A.S. Thom, 'Stonehenge' in *Journal for the History of Astronomy*, 5, 1974, 71–90

Thom, A., and A.S. Thom, 'Stonehenge as a lunar observatory' in *Journal for the History of Astronomy*, 6, 1975, 19–30

White, T., 'Human origins' in *New Scientist*, 16 November 2010, Supplement, i–viii

References

Williams, T.I., *A History of Inventions*, Time Warner Books, London, 2003

Wrangham, R., *Catching Fire: How Cooking Made Us Human*, Profile Books, London, 2009

Yong, E., 'Our hybrid origins' in *New Scientist*, 30 July 2011, 35–38

THE PURSUIT OF POWER

Bloom, A., 'Early probings, principles and practices' and 'Stationary engines' in *250 Years of Steam*, World's Work Ltd, Windmill Press, Kingswood, 1981, 6–16, 17–31

Bragg, M., 'Patent specification for Arkwright's spinning machine 1796' in *12 Books That Changed The World*, Hodder & Stoughton, London, 2006, 235–260

Clarke, D. (ed.), 'Newcomen's steam engine', 'Watt's steam engine' and 'Trevithick's steam engine' in *The Encyclopedia of Inventions*, Marshall Cavendish Books Ltd, London, 1977, 116–117, 118–119, 120–121

Crump, T., 'The atmospheric steam engine' and 'Cotton and wool: Lancaster and York' in *A Brief History of How the Industrial Revolution Changed the World*, Robinson Ltd, London, 2010, 23–54, 79–109

Crump, T., 'The invention of the steam engine and the problem of locomotion' in *A Brief History of the Age of Steam*, Carroll & Graf Publishers, New York, 2007, 50–70

Engels, F., *The Condition of the Working Class in England in 1846*, Basil Blackwell, Oxford, 1958

Feldman, A., and P. Ford, 'Sir Richard Arkwright' and 'James Watt' in *Scientists and Inventors*, Aldus Books Ltd, London, 1979, 62–63, 66–67

Macinnis, P., 'Discovering Steam Power' in *100 Discoveries*, Pier 9, Murdoch Books UK, London, 2009, 101–103

McKie, R., 'Where Watt worked his wizardry' in *The New Review* in *The Observer*, 4 September 2011, 21

Newhouse, E.L. (ed.), 'The power of steam' in *Inventors and Discoverers: Changing Our World*, National Geographic Society, Washington, DC, 1988, 14–43

Tames, R., 'In factory and foundry' in *Life during the Industrial Revolution*, Reader's Digest Association Ltd, London, 1995, 16–39

Trevelyan, G.M., *English Social History*, 3rd edn, Longmans, Green & Co., London, 1946

FULL STEAM AHEAD

Bloom, A., 'Steam pioneers', in *250 Years of Steam*, World's Work Ltd, Windmill Press, Kingswood, 1981, 178–184

Bridgeman, R., 'Trevithick and the Stephensons' in *1000 Inventions*, Dorling Kindersley, London, 2002, 26–35

Clarke, D. (ed.), 'Trevithick's steam engine' in *The Encyclopedia of Inventions*, Marshall Cavendish Books Ltd, London, 1977, 120–121

Crump, T., 'Transport on land and its limitations' and 'British railway mania' in *A Brief History of the Age of Steam*, Carroll & Graf Publishers, New York, 2007, 18–36, 147–185

Crump, T., 'Riverboats and railways' in *A Brief History of How the Industrial Revolution Changed the World*, Constable & Robinson Ltd, London, 2010, 169–197

Feldman, A., and P. Ford, 'George Stephenson' in *Scientists and Inventors*, Aldus Books, Middlesex, 1979, 110–111

Magnusson, M. (ed.), 'Transport' in *Reader's Digest Book of Facts*, 2nd edn, Reader's Digest, Auckland, 1995, 258–272

Newhouse, E.L. (ed.), 'On wheels and wings' in *Inventors and Discoverers: Changing Our World*, National Geographic Society, Washington, DC, 1988, 80–117

Norton, T., 'High, fast and hazardous' in *Smoking Ears and Screaming Teeth*, Arrow Books, London, 2010

Ransome, P.J.G., 'Rail' in I. McNeil (ed.), *An Encyclopedia of the History of Technology*, Routledge, Abingdon, 1990, 556–565

Strandh, S., 'Steam and combustion engines' in *The History of the Machine*, Bracken Books, London, 1989, 113–152

Sweet, M., *Inventing the Victorians*, Faber & Faber, London, 2001

Tames, R., 'A world transformed' in *Life during the Industrial Revolution*, Reader's Digest Association Ltd, London, 1995, 6–14

Taylor, G.R., 'Train' in *Inventions That Changed the World*, Reader's Digest Association Ltd, London, 1982, 303–306

Wolmar, C., *Fire and Steam*, Atlantic Books, London, 2007

LIKE A RED FLAG TO A BULL

Bloom, A., 'Steam for Agriculture' and 'Steam carriages and motor cars' in *250 Years of Steam*, World's Work Ltd, Windmill Press, Kingswood, 1981, 97–101, 135–145

Crump, T., 'The invention of the steam engine and the problem of locomotion' in *A Brief History of the Age of Steam*, Carroll & Graf Publishers, New York, 2007, 50–70

Feldman, A., and P. Ford, *Scientists and Inventors*, Aldus Books, Middlesex, 1979

Ford, H., with S. Crowther, *My Life and Work*, new edn, William Heinemann Ltd, London, 1924

Harmsworth, A.C., *Motors and Motor Driving*, Longmans, Green & Co., London, 1904

Hart-Davis, A., 'Tragedy of transport' in *The World's Stupidest Inventions*, Michael O'Mara Books Ltd, London, 2003, 19–22

Henderson, L. (ed.), *The Ultimate Motorcycle Book*, Dorling Kindersley, London, 1993

Hiscock, G.D., *Horseless Vehicles Automobiles and Motor Cycles*, Norman W. Henley & Co., New York, 1901

Homans, J.E., *Self-Propelled Vehicles*, 2nd edn, Theo. Audel & Co., New York, 1905

Macinnis, P., 'Inventing the internal combustion engine' in *100 Discoveries*, Pier 9, Murdoch Books UK Ltd, London, 2009, 174–175

Magnusson, M. (ed.), 'Transport' in *Reader's Digest Book of Facts*, Reader's Digest Association Ltd, London, 1994, 258–272

Mercredy, R.J. (ed.), *The Encyclopaedia of Motoring*, 3rd edn, Mercredy, Percy & Co. Ltd, Dublin, and Iliffe & Sons, Coventry and London, n.d. (circa 1906)

Newhouse, E.L. (ed.), 'On wheels and wings' in *Inventors and Discoverers: Changing Our World*, National Geographic Society, Washington, DC, 1998, 80–117

Sackville West, R., and M. Rickaby (eds), 'Conquering distance' in *Inventions That Changed the World*, Reader's Digest Association Ltd, London, 1997, 93–120

Taylor, G.R., 'Motor car' in *Inventions That Changed the World*, Reader's Digest Association Ltd, London, 1982, 179–184

Van Dulken, S., 'The automobile' and 'The internal combustion engine' [Otto] in *Inventing the 19th Century*, British Library, London, 2001, 26–27, 104–105

Van Dulken, S., 'Tarmac' [Hooley's patent] in *Inventing the 20th Century*, British Library, London, 2002, 32–33

Wallis Tayler, A.J., *Motor Cars or Power Carriages for Common Roads*, Crosby Lockwood & Son, London, 1897

Wise, D.B., *The Illustrated Encyclopedia of Automobiles*, New Burlington Books, London, 1979

Wolmar C., *Fire and Steam*, Atlantic Books, London, 2007

RUBBER DUB DUB

Clarke, D. (ed.), 'The wheel' in *The Encyclopedia of Inventions*, Marshall Cavendish Books Ltd, London, 1977, 8–9

De Bono, E., 'Pneumatic tyre' in *Eureka!*, Thames & Hudson, London, 1974, 34

Fawcett, P.H., *Exploration Fawcett*, The Companion Book Club, London, 1954

Grandin, G., *Fordlandia*, Icon Books Ltd, London, 2010

Ikenson, B., 'Vulcanization of Rubber' in *Patents: Ingenious Inventions*, Black Dog & Leventhal Publishers Inc., New York, 2004, 60–61

James, P., and N. Thorpe, 'Transportation' in *Ancient Inventions*, Michael O'Mara Books Ltd, London, 1995, 48–111

Macinnis, P., 'Gutta percha' in *100 Discoveries*, Pier 9, Murdoch Books UK Ltd, London, 2009, 144–146

Magnusson, M. (ed.), 'Transport' in *Reader's Digest Book of Facts*, Reader's Digest Association Ltd, London, 1994, 258–272

Mathé, J., *Leonardo's Inventions*, Liber SA, Fribourg-Genève, 1986

Newhouse, E.L. (ed.), 'Charles Goodyear' in *Inventors and Discoverers:*

Changing Our World, National Geographic Society, Washington, DC, 1988, 122–124

Taylor, G.R., 'Bicycle' in *Inventions That Changed the World*, Reader's Digest Association Ltd, London, 1982, 40–41

Van Dulken, S., 'The pneumatic tyre' [Dunlop's patent] and 'Vulcanised rubber' [Goodyear] in *Inventing the 19th Century*, British Library, London, 2001, 148–149, 206–207

MESSING ABOUT IN BOATS

Bloom, A., 'Marine engines' and 'Steam Pioneers' in *250 Years of Steam*, World's Work Ltd, Windmill Press, Kingswood, 1981, 33–63, 178–189

Bridgeman, R., 'John Wilkinson' in *1000 Inventions*, Dorling Kindersley, London, 2002, 109, 112

Crump, T., 'The invention of the steam engine' and 'Steam conquers the oceans' in *A Brief History of the Age of Steam*, Carroll & Graf Publishers, New York, 2007, 50–70, 284–320

Crump, T., 'Riverboats and railways' in *A Brief History of How the Industrial Revolution Changed the World*, Constable & Robinson Ltd, London, 2010 169–197

Editorial, 'Seafaring ancestor found in Philippines' in *New Scientist*, 12 June 2010, 16

Feldman, A., and P. Ford, 'Isambard Kingdom Brunel' in *Scientists and Inventors*, Aldus Books, Middlesex, 1979, 136–137

Heyerdahl, T., *The Kon Tiki Expedition*, George Allen & Unwin, London, 1950

Heyerdahl, T., *The Ra Expedition*, George Allen & Unwin, London, 1971

James, P., and N. Thorpe, 'Transportation' in *Ancient Inventions*, Michael O'Mara Books Ltd, London, 1995, 48–111

Magnusson, M. (ed.), 'Transport' in *Reader's Digest Book of Facts*, Reader's Digest Association Ltd, London, 1994, 258–272

Newhouse, E.L. (ed.), 'The power of steam' in *Inventors and Discoverers: Changing Our World*, National Geographic Society, Washington, DC, 1988, 14–43

Osborne, B.D., *The Ingenious Mr Bell*, Argyll Publishing, Glendaruel, 1995

Pain, S., 'When men were gods' in *New Scientist*, 10 February 2007, 46–47

Severin, T., *The Brendan Voyage*, Book Club Associates, London, 1978

Severin, T., *The China Voyage*, Little, Brown & Co., London, 1994

Throckmorton, P. (ed.), 'An Athenian warship recreated' in *The Sea Remembers*, Chancellor Press, London, 1996

Thucydides, *History of the Peloponnesian War*, Guild Publishing, London, 1990

Williams, T.I., 'New modes of transport' in *A History of Invention*, Time Warner Books UK, London, 97–108

IRONCLADS, DREADNOUGHTS AND SUBMERSIBLES

Anon. and M. Edwards, 'A Drebellian success' in *Historical Diving Times*, Summer 2003, 10–14

Boyle, R., *New Experiments, Physico-Mechanical, Touching the Spring of Air*, 1660, 363–365

Butler, A., 'Brunette sinks battleship' in *New Scientist*, 2 January 1999, 78–79

Cousteau, J.Y., and J. Dugan (eds), 'Letter to Thomas Jefferson' and 'The Admiralty secret circular' in *Underwater Treasury*, Hamish Hamilton, London, 1960, 15–16, 119–122

Chant, C., 'The submarine in World War 1' in *Submarines of the 20th Century*, Tiger Books International, London, 1966, 12–38

Collins, P., 'A ram for the rebels' in *New Scientist*, 4 January 2003, 44–45

Cook, C., and J. Stevenson, 'Naval warfare' in *Weapons of War*, Artus Publishing Co. Ltd, London, 1980, 102–116

Davies, R., 'The early pioneers' and 'The race to be first' in *Nautilus: The Story of Man Under the Sea*, BBC Books, 1995, 31–40, 41–63

Davis, R.H., 'The submarine boat' in *Deep Diving and Submarine Operations*, St Catherine Press, London, 1935, 486–502

Diolé, P., *Underwater Exploration: A History*, Elek, London, 1954

Dugan, J., 'The intelligent whale' in *Man Explores the Sea*, Penguin Books, Harmondsworth, 1966, 109–146

References

Ellis, R., 'Exploring the deep' in *Deep Atlantic*, Alfred A. Knopf, New York, 1996, 51–109

Feldman, A., and P. Ford, 'William Parsons' in *Scientists and Inventors*, Aldus Books, Middlesex, 1979, 220–221

Gould, D. (ed.), 'Clash of the Iron Ships' in *How Was It Done?*, Reader's Digest Association Ltd, London, 1995, 356–357

Keats, J., 'Transmission vamp' in *New Scientist*, 3 December 2011, 54

Kemp, P., 'Machines for the annoyance of shipping' in *Underwater Warriors*, Brockhampton Press, London, 1999, 11–21

Nicholson, A., *Men of Honour*, HarperCollins, London, 2005

Parsons, W.B., *Robert Fulton and the Submarine* [1922], AMS Press, New York, 1967

Rhodes, R., *Hedy's Folly: The Life and Breakthrough Inventions of Hedy Lamarr*, Doubleday, London, 2011

Tall, J.J., and P. Kemp, 'The first submersibles' in *HM Submarines in Camera*, Blitz Editions, Leicester, 1998, 1–40

Van Der Vat, D., 'Inventions and devices' in *Stealth at Sea*, Weidenfeld & Nicolson, London, 1994, 5–52

Van Dulken, S., 'The Resurgam submarine' [Garrett's patent] in *Inventing the 19th Century*, British Library, London, 2001, 160–161

Wilkins, J., 'Concerning the Possibility of training an Ark for Submarine Navigation; the Difficulties and Consequences of Such a Contrivance' in *Mathematical Magick*, S. Gellibrand, London, 1648

MORE COMPLICATED WAYS TO DROWN

Anon., 'Rapport sur une cloche ... plongeur inventée par M. Guillaumet' in *Comptes Rendus Hebdomadaires des Séances de l'Académie des Sciences*, 9, 1839, 363–366

Babbage, C., *Passages from the Life of a Philosopher*, Dawsons of Pall Mall, London, 1968

Bachrach, A.J., 'The history of the diving bell' in *Historical Diving Times*, 21, Spring 1998, 4–10

Bauer, J., 'Beyond bells – Halley's diving helmet' in *International Journal of Diving History*, 2 November 2006, 20–35

References

Bevan, J., 'The invention and development of the diving helmet and dress' in *Underwater Technology*, 17, Spring 1991, 19–25

Bevan, J., 'In defence of the Deanes' in *Historical Diving Society Newsletter*, 12 April 1995, 6–7

Bevan, J., *The Infernal Diver*, privately published, 1996 (copies available from the Historical Diving Society)

Bowman, G., *The Man Who Bought a Navy*, reprint, S. Birchall and P. Rowlands, Guildford, 1964

Cousteau, J.Y, and F. Dumas, 'Menfish' in *The Silent World*, Hamish Hamilton, London, 1953, 1–13

Davis, R.H., *Deep Diving and Submarine Operations*, St Catherine Press, London, 1935

Dean, C.A., *Submarine Researches* [1836], The Historical Diving Society, London, 2001

De Groot, J., 'The Sorima story' in *Historical Diving Times*, 43, Spring 2008, 26–33

De Latil, P., and J. Rivoire, 'The diving suit on its way' in *Man and the Underwater World*, Jarrolds, London, 1956, 141–177

Dick, P., 'The recent introduction' in *Historical Diving Times*, 28, Winter 2000, 29–33

Dugan, J., 'Treasure or trash' in *Man Explores the Sea*, Penguin Books, Harmondsworth, 1960, 78–108

Fardell, J., *John Lethbridge*, Historical Diving Society Monograph, 2011

Halley, E., 'The Art of Living under water: or, a Discourse concerning the Means of furnishing Air at the Bottom of the Sea . . .' in *Philosophical Transactions of the Royal Society*, 349, September 1716, 492–499

Halley, E., 'Addition to the art of living under water' in *Philosophical Transactions of the Royal Society*, 354, 1721, 177–180

Hamer, M., 'The doomsday wreck' in *New Scientist*, 21 August 2004, 36–38

Harris, J., 'Like a hog in a ditch' and 'Mutiny' in *Lost at Sea*, Guild Publishing, London, 1990, 9–40, 125–178

Jasinski, M., 'Meeting John Lethbridge' in *Historical Diving Times*, 46, Spring 2009, 12–17

Jung, M., 'Theodore Guillaumet's trials with his demand regulator' in *Historical Diving Times*, 23, Winter 1998, 14–15

Lowth, C.F., 'The Roman wrecks of Lake Nemi' in *Historical Diving Times*, 48, Winter 2010, 18–23

Madsen, A., 'The aqualung' in *Cousteau: An Unauthorised Biography*, Robson Books, London, 1989, 35–50

Masters, D., 'Seeking the Egypt's gold' and 'Salvage miracles at Scapa' in *When Ships Go Down*, Eyre & Spottiswood Ltd, London, 1945, 229–248, 271–289

Masters, D., 'Raising the Leonardo da Vinci battleship', reprinted in J.Y. Cousteau and J. Dugan, *Underwater Treasury*, Hamish Hamilton, London, 1960, 242–248

Pain, S., 'Saved from the bell' in *New Scientist*, 21 June 2003, 52–53

Pain, S., 'Secrets of the SS Persia' in *New Scientist*, 16 December 2006, 32–35

Scott, D., *Seventy Fathoms Deep*, Faber & Faber, London, 1931

Scott, D., *The Egypt's Gold*, Penguin Books, Harmondsworth, 1939

Wright, J., 'Egypt' in *Encyclopedia of Sunken Treasure*, Michael O'Mara Books Ltd, London, 1995, 224–229

ON A WING AND A PRAYER

Acton, J.A., *The Man Who Touched the Sky*, Sceptre, London, 2002

Bagley, J.A., 'Aeronautics' in I. McNeil (ed.), *An Encyclopedia of the History of Technology*, Routledge, Abingdon, 1990, 609–647

Cayley, G., 'On aerial navigation' in *Journal of Natural Philosophy, Chemistry and the Arts*, 24, 1809, 164–174; 25, 1810, 81–87, 161–169

Chant, C., *Aviation*, Orbis Publishing, London, 1978, 9–71

Churchill, W.S., *The Great War*, vol. 1, George Newnes Ltd, London, n.d.

Crouch, T.D., C. Hart, P. Marvelas, and D.A. Pisano (eds), *The Genesis of Flight*, The Friends of the US Airforce Academy Library and University of Washington Press, 2000

Frater, A., *The Balloon Factory*, Picador, London, 2008

Garber, L.W., *The Wright Brothers and the Birth of Aviation*, The Crowood Press, Marlborough, 2005

References

Gibbs-Smith, C.H., *Sir George Cayley's Aeronautics 1796–1855*, HMSO, London, 1962

Haldane, J.B.S., 'On being the right size' in *Possible Worlds*, Chatto & Windus, London, 1945, 18–26

Holmes, R., *The Age of Wonder*, HarperPress, London, 2008

Ikenson, B., 'Airplane' and 'Zeppelin' in *Patents: Ingenious Inventions*, Black Dog & Leventhal Publishers Inc., New York, 2004, 14–16, 183–185

James, P., and N. Thorpe, 'Man-bearing kites and parachutes' in *Ancient Inventions*, Michael O'Mara Books Ltd, London, 1995, 104–107

Jarrett, P. (ed.), *Pioneer Aircraft*, Putnam Aeronautical Books, London, 2002

Kelly, M., *Steam in the Air*, Pen & Sword Aviation, Barnsley, 2006

Lilienthal, O., *Bird Flight as a Basis of Aviation*, Longmans, London, 1911

Mondey, D. (ed.), *The International Encyclopedia of Aviation*, Octopus Books Ltd, London, 1977

Penrose, H., *An Ancient Air*, An Airlife Classic, Shrewsbury, 2000

Pollack, K., and E.A. Underwood, *The Healers*, Thomas Nelson & Sons, London, 2002

Sackville West, R., and M. Rickaby (eds), 'Conquering distance' in *Inventions That Changed the World*, Reader's Digest Association Ltd, London, 1997, 93–120

Sweeny, C., 'Alchemist dismissed as a crank may have been the first man to fly, academic claims' in *The Times*, 8 September 2008, 4

Van Dulken, S., 'The aeroplane' [Wright brothers' patent] in *Inventing the 20th Century*, British Library, London, 2002, 18–19

Wells, H.G., *Anticipations*, Harper, New York, 2002

Wilkins, J., *A Discourse Concerning a New World and Another Planet*, John Maynard, London, 1640

SAFETY FIRST

Acton, J.A., 'Excelsior' in *The Man Who Touched the Sky*, Sceptre, London, 2002, 105–127

References

Anon., 'Robert's Safety Hood', Whitehaven Museum Information Sheet, Friends of Whitehaven Museum, 1986, 3pp.

Chance, T., and P. Williams, *Lighthouses: The Race to Illuminate the World*, New Holland Publishers (UK) Ltd, London, 2008

Chant, C., 'Introduction' in *Aviation*, Orbis Publishing, London, 1978, 9–23

Crump, T., 'Signalling, safety and traffic control' in *A Brief History of the Age of Steam*, Carroll & Graf Publishers, New York, 2007, 167–179

Dale, R., and J. Gray, *Edwardian Inventions 1901–1905*, Star Books, W.H. Allen, London, 1979

Faith, N., *Crash*, Boxtree, London, 1997

Gale, A., 'Watchers and waiters: regulating the coast' in *Britain's Historic Coast*, Tempus Publishing, Stroud, 2000, 115–125

Hart-Davies, A., 'A wonder of the world' in *What the Past Did for Us*, BBC Books, London, 2004, 134–135

Hope, E., 'Lighthouse homes' in *Grace Darling: Heroine of the Farne Islands*, Walter Scott Ltd, London, 61–87

Howard, P., 'Aerospace medicine' in Mondey, D. (ed.), *The International Encyclopedia of Aviation*, Octopus Books Ltd, London, 1977, 138–145

Ikenson, B., 'Airbags', 'Parachute', 'Seat belt' and 'Towering inferno' in *Patents: Ingenious Inventions*, Black Dog & Leventhal Publishers Inc., New York, 2004, 112–114, 135–137, 143–144, 181–182

James, P., and N. Thorpe, 'Lighthouse' and 'Fire engines' in *Ancient Inventions*, Michael O'Mara Books Ltd, London, 1995, 98–101, 368–373

Middleton, D., 'Britain enters the jet age' in *Test Pilots*, Guild Publishing, London, 124–137

Norton, T., 'Seeing the light' in *Under Water to Get Out of the Rain*, Arrow Books, London, 2004, 1–9

Norton, T., 'Adrift and alone' and 'High, fast and hazardous' in *Smoking Ears and Screaming Teeth*, Arrow Books, London, 2011, 255–274, 321–343

Palmer, M., *Eddystone: The Finger of Light*, Seafarer Books, Woodbridge, 2005

Paulin, G., 'Appareil pour éteindre les feux de cave' in *Annales Maritimes et Coloniales*, 1, 2nd series, 1835

Pepys, S., *The Diary of Samuel Pepys 1660–1669*, G. Bell & Sons, London, 1926, 330–337

Ranson, P.J.G., 'The railway at work' and 'Signalling' in I. McNeil (ed.), *An Encyclopedia of the History of Technology*, Routledge, Abingdon, 1990, 562–565, 592

Taylor, G.R., 'Fire Engines' in *Inventions That Changed the World*, Reader's Digest Association Ltd, London, 1982, 113

Van Dulken, S., 'The cork-filled life preserver' [Guerin's patent] and 'The parachute fire escape' [Oppenheimer] in *Inventing the 19th Century*, British Library, London, 2001, 66–67, 136–137

Wolmar, C., 'Getting the railway habit' and 'Danger and exploitation on the tracks' in *Fire and Steam*, Atlantic Books, London, 2007, 43–59, 146–163

SEEING THE LIGHT

Bowers, B., 'Electricity' in I. McNeil (ed.), *An Encyclopedia of the History of Technology*, Routledge, Abingdon, 350–387

Bragg, M., 'Faraday's Experimental researches in electricity' in *12 Books That Changed The World*, Hodder & Stoughton, London, 2006, 207–234

Dale, R., and J. Gray, *Edwardian Inventions 1901–1905*, Star Books, W.H. Allen, London, 1979

Fellowes-Gordon, I., 'Thomas Alva Edison' in J. Canning (ed.), *100 Great Lives*, Odhams Books Ltd, London, 1965, 192–196

Franklin, B., 'A letter of Benjamin Franklin Esq. to Mr Peter Collinson F.R.S. concerning an electrical kite' in *Philosophical Transactions of the Royal Society*, 47, 1751–1752, 565–567

Gould, D. (ed.), 'The magic workshop of a modern wizard' in *How Was it Done?*, Reader's Digest Association Ltd, London, 1995, 278–281

Green, M., 'Rogue elements' in *The Nearly Men*, Tempus Publishing, Stroud, 2007, 128–152

Hirshfeld, A., *The Electric Life of Michael Faraday*, Raincoast Books, Vancouver, 2006

References

Ikenson, B., 'Lightbulb' and 'Battery' in *Patents: Ingenious Inventions*, Black Dog & Leventhal Publishers Inc., New York, 2004, 32–35, 73–76

James, F., 'Electric battery' in P. Tallack (ed.), *The Science Book*, Weidenfeld & Nicolson, London, 2001, 116–117

James, P., and N. Thorpe, 'Electric batteries' in *Ancient Inventions*, Michael O'Mara Books Ltd, London, 1995, 145–152

Levy, J., 'Tesla v Edison' in *Scientific Feuds*, New Holland Publishers (UK) Ltd, London, 2010, 196–203

Knight, D., 'New elements' in P. Tallack (ed.), *The Science Book*, Weidenfeld & Nicolson, London, 2001, 122–123

Macinnis, P., 'Inventing the electric cell', 'Inventing electric light', 'The first electric motor' and 'Harnessing alternating current' in *100 Discoveries*, Pier 9, Murdoch Books UK Ltd, London, 2009, 116–118, 119–1200, 131–133, 191–193

McNicol, T., *AC/DC: The Savage Tale of the First Standards War*, John Wiley & Sons, New York, 2006

Meyer, H.W., *A History of Electricity and Magnetism*, Burndy Library, Norwalk, 1972

Pollard, J., 'The great adventure' in *Boffinology*, John Murray, London, 2010, 4–39

Tucker, T., *Bolt of Fate: Benjamin Franklin and His Electric Kite Hoax*, Sutton Publishing Ltd, Stroud, 2004

Spangenburg, R., and D.K. Moser, 'Atoms and elements' and 'Magnetism, electricity and light' in *The Age of Synthesis 1800–1895*, Facts on File Inc., New York, 2004, 3–21, 46–64

Van Dulken, S., 'The alternating current induction motor' [Tesla's patent] in *Inventing the 19th Century*, British Library, London, 2001, 14–15

ON LINE

Bridgeman, R., 'Morse and Vail' in *1000 Inventions*, Dorling Kindersley, London, 2002, 130–131

De Bono, E. (ed.) 'Electric telegraph' in *Eureka!*, Thames & Hudson, London, 1974

Feldman, A., and P. Ford, 'Samuel Morse' in *Scientists and Inventors*, Aldus Books, Middlesex, 1979, 122–123

References

Forde-Johnston, J., *Hadrian's Wall*, Book Club Associates, London, 1977

Gooch, D., *The Diaries of Daniel Gooch* [1892], Nonsuch Publishing Ltd, Stroud, 2006

Gould, D., 'Linking two continents with a single cable' in *How Was It Done?*, Reader's Digest Association Ltd, London, 1995, 392

Green, M., 'Rogue elements' in *The Nearly Men*, Tempus Publishing, Stroud, 2007, 128–152

Ikenson, B., 'Telegraph' in *Patents: Ingenious Inventions*, Black Dog & Leventhal Publishers Inc., New York, 2004, 53–55

Macinnis, P., 'Inventing the telegraph' in *100 Discoveries*, Pier 9, Murdoch Books UK Ltd, London, 2009, 104–105

Maury, M.F., 'The depths of the ocean' and 'The basin of the Atlantic' in *The Physical Geography of the Sea*, Thomas Nelson & Sons, London, 6th edn, 1893, 292–305, 306–325

Merret, J., *Three Miles Down: The Story of the Transatlantic Cable*, Hamish Hamilton, London, 1958

Meyer, H.W., 'The electric telegraph' in *A History of Electricity and Magnetism*, Burndy Library, Norwalk, 1972, 95–130

Ohlman, H., 'The telegraph' in I. McNeil (ed.), *An Encyclopedia of the History of Technology*, Routledge, Abingdon, 1990, 710–727

Pollard, J., 'Physics, Football and Fear' in *Boffinology*, John Murray, London, 2010, 41–69

Ranson, P.J.G., 'The telegraph' in I. McNeil (ed.), *An Encyclopedia of the History of Technology*, Routledge, Abingdon, 1990, 710–717

Thompson, D., *England in the Nineteenth Century*, Penguin Books, Harmondsworth, 1964

Van Dulken, S., 'The electric telegraph' [Cooke and Wheatstone's patent] in *Inventing the 19th Century*, British Library, London, 2001, 82–83, 192–196

WIRED FOR SOUND

Bridgeman, R., 'Wiring the world' and 'Doing away with distance' in *1000 Inventions*, Dorling Kindersley, London, 2002, 130–131, 152–153

References

Bryson, B., 'The passage' in *At Home*, Black Swan, London, 2010, 307–340

Feldman, A., and P. Ford, 'Alexander Graham Bell' in *Scientists and Inventors*, Aldus Books, Middlesex, 1979, 204–205

Gould, D., 'Electricity harnessed to carry spoken words' in *How Was It Done?*, Reader's Digest Association Ltd, London, 1995, 394–395

Green, M., 'Crossed lines' in *The Nearly Men*, Tempus Publishing, Stroud, 2007, 13–37

Ikenson, B., 'Telephone' in *Patents: Ingenious Inventions*, Black Dog & Leventhal Publishers Inc., New York, 2004, 55–57

Macinnis, P., 'Inventing the telephone' in *100 Discoveries*, Pier 9, Murdoch Books UK Ltd, London, 2009, 186–188

Magnusson, M. (ed.), 'Communications linking the world' in *Reader's Digest Book of Facts*, Reader's Digest Association Ltd, London, 1994, 250–251

Meyer, H.W., 'The telephone' in *A History of Electricity and Magnetism*, Burndy Library, Norwalk, 1972, 131–151

Ohlman, H., 'Information' in I. McNeil (ed.), *An Encyclopedia of the History of Technology*, Routledge, Abingdon, 1990, 686–758

Sackville West, R., and M. Rickaby (eds), 'The whole world at your fingertips' in *Inventions That Changed the World*, Reader's Digest Association Ltd, London, 1997, 56–57

Taylor, G.R., 'Telephone' in *Inventions That Changed the World*, Reader's Digest Association Ltd, London, 1982, 288–291

Van Dulken, S., 'The automatic telephone exchange' [Strowger's patent] and 'The telephone' [Bell] in *Inventing the 19th Century*, British Library, London, 2001, 24–25, 190–191

ON THE CREST OF A WAVE

Anarnthaswamy, A., 'Hip, hip, array' in *New Scientist*, 24 March 2012, 43–47

Aughton, P., 'Beyond the visible spectrum' in *The Story of Astronomy*, Quercus, London, 2011, 158–167

Bridgeman, R., 'Doing away with distance' and 'Seeing with radio' in *1000 Inventions*, Dorling Kindersley, London, 2002, 152–153, 198–199

References

Buchanan, N. (ed.) 'How aircraft and ships see with radar' in *How Is It Done?*, Reader's Digest Association Ltd, London, 1990, 154–155

Clarke, D. (ed.), 'Radar' in *The Encyclopedia of Inventions*, Marshall Cavendish Books Ltd, London, 1977, 144–145

De Bono, E. (ed.), 'Sonar', 'Radio Telescope' and 'Radar' in *Eureka!*, Thames & Hudson Ltd, London, 1974, 199–200, 201, 202

Feldman, A., and P. Ford, 'Sir Robert Watson-Watt' in *Scientists and Inventors*, Aldus Books, Middlesex, 1979, 284–285

Gould, D., 'The spark that set the world talking' in *How Was It Done?*, Reader's Digest Association Ltd, London, 1995, 198–199

Green, M., 'Making waves' in *The Nearly Men*, Tempus Publishing, Stroud, 2007, 81–127

Hambling, D., 'Airborne radar will map the ground in 3D' in *New Scientist*, 2 April 2011, 24

Ikenson, B., 'Radio' [Tesla's patent] in *Patents: Ingenious Inventions*, Black Dog & Leventhal Publishers Inc., New York, 2004, 254–257

Jessen, C., 'The social message of "text neck"' in *Evening Standard*, 12 October 2011, 37

Jha, A., 'Space mysteries glimpsed from on top of the world' in *The New Review* in *The Observer*, 29 January 2012, 18–19

Larson, E., *Thunderstruck*, Doubleday, London, 2006

Macinnis, P., 'Discovering the thermionic effect' and 'Discovering radio waves' in *100 Discoveries*, Pier 9, Murdoch Books UK Ltd, London, 2009, 194–198, 201–203

Magnusson, M. (ed.), 'Weapons and warfare' in *Reader's Digest Book of Facts*, Reader's Digest Association Ltd, London, 1994, 237–239

Marks, P., 'The gentleman hacker' in *New Scientist*, 24–31 December 2011, 48–49

Meyer, H.W., 'Microwaves' in *A History of Electricity and Magnetism*, Burndy Library, Norwalk, 1972, 253–260

Ohlman, H., 'Information' in I. McNeil (ed.), *An Encyclopedia of the History of Technology*, Routledge, Abingdon, 1990, 686–758

Pain, S., 'Pinnipeds on parade' in *New Scientist*, 14 February 2004, 48–49

Pain, S., 'Two men and a wheelbarrow' in *Farmer Buckley's Exploding Trousers*, Profile Books, London, 2011, 172–176

Pukas, A., 'Is this the cold war's most bizarre untold story?' in *Daily Express*, 9 August 2012, 30–31

Sackville West, R., and M. Rickaby (eds), "Radio rings the world' and 'The whole world at your fingertips' in *Inventions That Changed the World*, Reader's Digest Association Ltd, London, 1997, 54–57

Taylor, G.R., 'Radar' and 'Radio telescope' in *Inventions That Changed the World*, Reader's Digest Association Ltd, London, 1982, 223–224

Van Der Vat, D., *Stealth at Sea*, Weidenfeld & Nicolson, London, 1994

Van Dulken, S., 'The radio' [Marconi's patent] in *Inventing the 19th Century*, British Library, London, 2001, 158–159

Van Dulken, S., 'Radar' [Watson-Watt's patent] and 'Microwave oven' [Spenser] in *Inventing the 20th Century*, British Library, London, 2002, 100–101, 116–117

CAPTURED IN TIME

BBC1 TV, *The Weird Adventures of Eadweard Muybridge*, 30 November 2010

Boucher, C., and C. Smith, 'Capturing a likeness' in *Extraordinary Origins of Everyday Things*, Reader's Digest Association Ltd, London, 2009, 192–193

Burke J., 'Faith in numbers' in *Connections*, Macmillan London Ltd, London, 1978, 81–113

Clark, W., 'Photography' in S.P. Parker (ed.), *Concise Encyclopedia of Science and Technology*, 3rd edn, McGraw-Hill, New York, 1994, 1410–1411

Conrad, P., 'The genius of movement who made the world stand still' in *The New Review* in *The Observer*, 29 August 2010, 24–25

De Bono, E. 'Cinematograph' in *Eureka!*, Thames & Hudson Ltd, London, 1974, 56–57

Feldman, A, and P. Ford, 'Johannes Gutenberg' and 'William Henry Fox Talbot' in *Scientists and Inventors*, Aldus Books, Middlesex, 1979, 16–17, 128–129

Gibbons-Fly, W., 'Cinematography' in S.P. Parker (ed.), *Concise Encyclopedia of Science and Technology*, 3rd edn, McGraw-Hill, New York, 1994, 393–394

References

Goff, A., 'Birth of digital snaps' in *New Scientist*, 10 March 2012, 28–29

Gould, D., 'The magic workshop of the modern wizard' in *How Was It Done?*, Reader's Digest Association Ltd, London, 1995, 278–281

Hope, A., 'A century of recorded sound' in *New Scientist*, 22–29 December 1977, 797–799

Hope, A., 'From rubber disc to magnetic tape' in *New Scientist*, 12 January 1978, 96–97

Ikenson, B., 'Camera' and 'Phonograph' in *Patents: Ingenious Inventions*, Black Dog & Leventhal Publishers Inc., New York, 2004, 18–20, 252–254

James, P., and N. Thorpe, 'Books and printing' in *Ancient Inventions*, Michael O'Mara Books Ltd, London, 1995, 512–517

Lindgren, E., *The Art of Film*, George Allen & Unwin Ltd, London, 1948

Macinnis, P., 'Inventing moveable type' and 'Inventing photography' in *100 Discoveries*, Pier 9, Murdoch Books UK Ltd, London, 2009, 61–63, 154–156

Magnusson, M. (ed.), 'Cinema and television' in *Reader's Digest Book of Facts*, Reader's Digest Association Ltd, London, 1994, 344–347

Maré, E. de, *Photography*, Penguin Books, Harmondsworth, 1957

Marks, P., 'Play that tape again' in *New Scientist*, 20 October 2012, 20

Morgan, R., and G. Perry (eds), *1000 Makers of the Cinema*, Times Newspapers Ltd, London, 1995

Naughton, J., 'The lesson we can learn from the rise and fall of Kodak' in *The New Review* in *The Observer*, 15 January 2012, 24–25

Ohlman, H., 'Communication and calculation' in I. McNeil (ed.), *An Encyclopedia of the History of Technology*, Routledge, Abingdon, 1990, 729–758

O'Neill, J.P., 'Camera' in S.P. Parker (ed.), *Concise Encyclopedia of Science and Technology*, 3rd edn, McGraw-Hill, New York, 1994, 312–313

Pollack, P., *The Picture History of Photography*, 2nd edn, Harry H. Abrams Inc., New York, 1970

Taylor, G.R., 'Motion pictures', 'Printing' and 'Record player' in *Inventions That Changed the World*, Reader's Digest Association Ltd, London, 1982, 173–178, 216–222, 230–233

Van Dulken, S., 'The motion picture camera' [Le Prince's Patent] and 'The phonograph' [Edison] in *Inventing the 20th Century*, British Library, London, 2002, 126–127, 142–143

Watson, R., and H. Rappaport, *Capturing Light: A True Story of Genius, Rivalry and the Birth of Photography*, Pan Macmillan, London, 2013

West, R.S., and M. Rickaby (eds), 'Sounds of the century' and 'The silver screen' in *Inventions That Changed the World*, Reader's Digest Association Ltd, London, 1997, 136–139

THE CAT'S WHISKERS

Barlow, G.H., 'Liquid crystals' in S.P. Parker (ed.), *Concise Encyclopedia of Science and Technology*, 3rd edn, McGraw-Hill, New York, 1994, 71

Bridgeman, R., 'Making dreams come true' in *1000 Inventions*, Dorling Kindersley, London, 2002, 202–203

Buchanan, N. (ed.), 'Remote controls operating switches from a distance' in *How Is It Done?*, Reader's Digest Association Ltd, London, 1990, 220

Clarke, D. (ed.), 'Radio and television' in *The Encyclopedia of Inventions*, Marshall Cavendish Books Ltd, London, 1977, 142–143

Cock, G., 'Looking forward. A personal forecast of the future of television' in *Radio Times*, 23 October 1936, 6–7

Collins, P., 'Live from the Paris Opera' in *New Scientist*, 12 January 2008, 44–45

De Bono, E. (ed.), 'Triode' and 'Television' in *Eureka!*, Thames & Hudson Ltd, London, 1974, 58–59, 59–61

De Santick, S., 'Television' in S.P. Parker (ed.), *Concise Encyclopedia of Science and Technology*, 3rd edn, McGraw-Hill, New York, 1994, 1883

Feldman, A., and P. Ford, 'John Logie Baird' in *Scientists and Inventors*, Aldus Books, Middlesex, 1979, 276–277

Ikenson, B., 'Remote control' [Adler's patent] and Television [Zworykin's patent] in *Patents: Ingenious Inventions*, Black Dog & Leventhal Publishers Inc., New York, 2004, 222–224, 268–271

Macinnis, P., 'Discovering the transistor' in *100 Discoveries*, Pier 9, Murdoch Books UK Ltd, London, 2009, 244–246

Magnusson, M. (ed.), 'Cinema and television' in *Reader's Digest Book of Facts*, Reader's Digest Association Ltd, London, 1994, 344–347

Meyer, H.W., 'Plasmas, masers, lasers, fuel cells, piezoelectric crystals, transistors' in *A History of Electricity and Magnetism*, Burndy Library, Norwalk, 1972, 275–298

Roach, J., *Armchair Nation: An Intimate History of Britain in Front of the TV*, Profile Books, London, 2013

Tannas Jr., L.E., 'Electronic display' in S.P. Parker (ed.), *Concise Encyclopedia of Science and Technology*, 3rd edn, McGraw-Hill, New York, 1994, 685–686

Taylor, G.R., 'Radio' and 'Television' in *Inventions That Changed the World*, Reader's Digest Association Ltd, London, 1982, 226–229, 292–297

Van Dulken, S., 'Television' [Baird's patent] and 'The transistor' [Bardeen and Brattain] in *Inventing the 20th Century*, British Library, London, 2002, 74–75, 122–123

West, R.S., and M. Rickaby (eds), 'Television broadcasting is born' in *Inventions That Changed the World*, Reader's Digest, London, 1997, 142–143

Williams, T.I., 'Telecommunications' in *A History of Invention*, Time Warner Books UK, London, 216–217

ALL THINGS GREAT AND SMALL

Aughton, P., *The Story of Astronomy*, Quercus, London, 2008

Defoe, D., *A Journal of the Plague Year* [1772], Penguin Books, Harmondsworth, 1970

Feldman, A., and P. Ford, 'Hans Lippershey' and 'Antoine van Leeuwenhoek' in *Scientists and Inventors*, Aldus Books, Middlesex, 1979, 28–29, 44–45

Holmes, R., 'Herschel on the moon' in *The Age of Wonder*, HarperPress, London, 2008, 60–124

Hooke, R., *Robert Hooke's Micrographia* [1665], Science Heritage Ltd, Lincolnwood, 1989

James, P., and N. Thorpe, 'Magnifying glasses' in *Ancient Inventions*, Michael O'Mara Books Ltd, London, 1995, 157–161

Jardine, L., 'Through a glass darkly' and 'Subtle anatomy' in *Ingenious Pursuits*, Little, Brown & Co., London, 1999, 42–50, 90–132

Levy, J., 'Galileo and Pope Urban' in *Scientific Feuds*, New Holland Publishers (UK) Ltd, London, 166–175

Macinnis, P., 'Inventing the telescope' and 'Inventing the microscope' in *100 Discoveries*, Pier 9, Murdoch Books UK Ltd, London, 2009, 71–73, 89–90

Merchant J., 'Decoding the Antikythera' in *New Scientist*, 13 December 2008, 36–40

Pain, S., 'Into the Breeches' in *New Scientist*, 28 May 2011, 37–39

Sagan, C., *Cosmos*, Macdonald Futura Publishers, London, 1980

Throckmorton, P., 'The road to Gelidonya' in *The Sea Remembers*, Chancellor Press, London, 1987, 14–23

THE GERM OF AN IDEA

Baxby, D., 'The end of smallpox' in *History Today*, March 1999, 14–16

BBC4 TV, *Breaking the mould*, 19 April 2011

Cheyne, W.W., 'Lord Lister 1827–1912' in *Obituary Notices of Fellows Deceased*, Royal Society, London, 1912, 615–621

Diggings, F.W.E., 'The discovery of penicillin' in *Biologist*, 47, June 2000, 115–119

Fisher, R.B., *Joseph Lister 1827–1912*, Macdonald & Jane's, London, 1977

Hollingham, R., *Blood and Guts: A History of Surgery*, BBC Books, London, 2008

Jenner, E., 'An inquiry into the causes and effects of the *variolae vaccinae* . . . known by the name of the cow pox' [1798], extract in L. Clendening, *Source Book of Medical History*, Dover Publications Inc., New York, 1942, 294–300

Koch, R., 'Investigations into the etiology of traumatic infectious diseases' [1878] in Classics of Medicine Library, The New Sydenham Society, 1880

Lister, J., 'The antiseptic system: on a new method of treating compound fracture, abscess etc., with observations on the conditions of suppuration' in *The Lancet*, 30 March 1867

Lister, J., 'On the effects of the antiseptic system of treatment upon the salubrity of a surgical ward' in *The Lancet*, 1 January 1870, 4–6

Lister, J., 'Presidential address' in *Report of the 66th meeting of the British Association of Science, Liverpool*, John Murray, London, 1896, 3–27

Longmate, N.R., *King Cholera: The Biography of a Disease*, Hamish Hamilton, London, 1966

Norton, T., 'The desire for disease' and 'Risky business' in *Smoking Ears and Screaming Teeth*, Arrow Books, 2011, 106–126, 344–353

Nuland, S.B., 'Joseph Lister's antiseptic surgery' in *Doctors: The Illustrated History of Medical Pioneers*, Black Dog & Leventhal Publishers Inc., New York, 2008, 324–365

Pain, S., 'The poet and the pox' in *New Scientist*, 18 January 2003, 44–45

Pain, S., 'To Russia with trepidation' in *New Scientist*, 24 March 2007, 56–57

Pasteur, L., 'Mémoires sur les corpuscules organisés qui existent dans l'atmosphère. Examen de la doctrine des générations spontanées' in *Annales des Sciences Naturelles*, Paris, 1862

Porter, R., 'From Pasteur to penicillin' in *The Greatest Benefit to Mankind: A Medical History of Humanity from Antiquity to the Present*, HarperCollins, London, 428–461

Simmonds, J.G., 'Louis Pasteur', 'Robert Koch', 'Edward Jenner' and 'Howard Florey' in *Doctors and Discoveries*, Mifflin Co., Boston and New York, 2002, 18–23, 24–28, 152–156, 245–250

Treves, F., 'The Old Receiving Room' in *The Elephant Man and Other Reminiscences*, Cassell & Co. Ltd, London, 1923, 39–58

Wainwright, M., 'Who witnessed the discovery of penicillin?' in *Biologist*, 52, August 2005, 234–237

References

A DIRTY BUSINESS

Ashenburg, K., *The Dirt on Clean*, North Point Press, New York, 2007

BBC4 TV, ap Glyn, Ifor, *The Toilet – An Unspoken History*, 26 November 2012

BBC2 TV, Snow, D., *Filthy Cities: Medieval London*, 5 April 2011

Bryson, B., 'The bathroom' in *At Home*, Black Swan, London, 2010, 486–526

Halliday, S., *The Great Filth*, Sutton Publishing, Stroud, 2007

Hart-Davis, A., 'The Romans' in *What the Past Did For Us*, BBC Books, London, 2004, 145–167

Hytch, E.J., 'Public Baths' in *The Lancet*, 8 August 1835, 204–206

Norton, T., 'Seeing the Light' in *Under Water to Get Out of the Rain*, Arrow Books, London, 2004, 1–9

Smythe, R., 'Wiping up' in *New Scientist*, 22–29 December 2012, 74–75

Snow, J., 'On the mode of communication of cholera', extract in L. Clendening, *Source Book of Medical History*, Dover Publications Inc., New York, 1942, 468–473

Van Dulken, S., 'The silent valveless waste water preventer' [Giblin's patent] in *Inventing the 19th Century*, British Library, London, 2001, 178–179

Worsley, L., *If Walls Could Talk: An Intimate History of the Home*, Faber & Faber, London, 2012

INSIDE INFORMATION

Ayer, W., 'Account of an eye-witness to the first public demonstration of ether anaesthesia at the Massachusetts General Hospital, October 16, 1846' in L. Clendening, *Source Book of Medical History*, Dover Publications Inc., New York, 1942, 372–373

Booth, F., H.J. Bigelow and R. Liston, 'Surgical operations performed during insensibility, produced by the inhalation of ether' in *The Lancet*, 2 January 1847, 5–8

Burnie, D., 'Imaging the body' in *Milestones in Medicine*, Reader's Digest Association Ltd, London, 1998, 21–29

References

Davy, H., *Researches, Chemical and Philosophical; Chiefly Concerning Nitrous Oxide, or Dephlogisticated Nitrous air, and its Respiration*, J. Johnson, London, 1800

Editorial, 'Administration of chloroform to the Queen' in *The Lancet*, 14 May 1853, 453

Editorial, 'Nobel in defeat' in *New Scientist*, 18 October 2003, 5

Hollingham, R., 'Bloody beginnings' in *Blood and Guts: A History of Surgery*, BBC Books, London, 2008, 31–101

Holmes, R., 'Davy on the gas' in *The Age of Wonder*, HarperPress, London, 2008, 235–304

Jones, R., and O. Lodge, 'The discovery of a bullet lost in the wrist by means of Roentgen ray' in *The Lancet*, 22 February 1896, 476–477

Laënnec, R.T.H., *Traité de l'Auscultation Médiate*, 2 vols., J.A. Brosson & J.S. Chaude, Paris, 2nd edn, 1826

Long, C.W., 'First surgical operation under ether' [1853] in L. Clendening, *Source Book of Medical History*, Dover Publications Inc., New York, 1942, 356–358

Macinnis, P., 'Discovering X-rays' in *100 Discoveries*, Pier 9, Murdoch Books UK Ltd, London, 2009, 208–210

Morton, W.T.G., 'Remarks on the proper mode of administering ether by inhalation' [1847] in L. Clendening, *Source Book of Medical History*, Dover Publications Inc., New York, 1942, 366–372

Norton, T., 'Sniff it and see', 'That unhealthy glow' and 'A change of heart' in *Smoking Ears and Screaming Teeth*, Arrow Books, 2011, 23–45, 145–169, 203–210

Nuland, S.B., 'René Laennec, inventor of the stethoscope' and 'Surgery without pain' in *Doctors: The Illustrated History of Medical Pioneers*, Black Dog & Leventhal Publishers Inc., New York, 2008, 192–227, 253–288

Porter, R., 'Scientific medicine in the 19th century' and 'Surgery after Lister' in *The Greatest Benefit to Mankind*, Fontana Press, London, 1999, 304–347, 597–627

Röntgen, W.C., 'Uber Eine Neue Art von Strahlen' [1895] in L. Clendening, *Source Book of Medical History*, Dover Publications Inc., New York, 1942, 666–675

Simmons, J.G., 'René Laënnec: The physician's new gaze', 'W.C. Röntgen: The discovery of X-rays' and 'Raymond Damadian' in *Doctors and Discoveries*, Houghton Mifflin Co., Boston and New York, 2002, 62–66, 102–104, 356–360

Simpson, J.Y., 'On a new anaesthetic agent, more efficient than sulphuric ether' in *The Lancet*, 20 November 1847, 549–550

Sistare Jnr., G., 'Nuclear magnetic resonance' in S.P. Parker (ed.), *Concise Encyclopedia of Science and Technology*, 3rd edn, McGraw-Hill, New York, 1994, 1270

Snow, S.J., *Blessed Days of Anaesthesia*, Oxford University Press, Oxford, 2008

Van Dulken, S., 'Computer-aided tomography' [Hounsfield's patent] and 'Magnetic resonance imaging' [Damadian] in *Inventing the 20th Century*, British Library, London, 2002, 150–151, 170–171

West, R.S., and M. Rickaby (eds), 'Aids to diagnosis' in *Inventions That Changed the World*, Reader's Digest Association Ltd, London, 1997, 68–69

BLOODY HELL

BBC2 TV, *Frontline Medicine*, 20 November 2011

Blundell, Dr, 'Observations on the transfusion of blood. With a description of his Gravitator' in *The Lancet*, 13 June 1829, 321–324

Burnie, D., 'Rapid response', 'The rise of modern surgery' and 'Technology and the body' in *Milestones in Medicine*, Reader's Digest Association Ltd, London, 1998, 35–36, 65–75, 76–85

Coghlan, A., 'Big red boost for artificial blood' in *New Scientist*, 23 August 2008, 10

Cohen, D., 'Code red' in *New Scientist*, 22 October 2011, 48–51

Editorial, 'Don't know your own blood group? Doesn't matter' in *New Scientist*, 7 April 2007, 8

Editorial, 'There will be blood. Fake blood' in *The New Review* in *The Observer*, 23 October 2011, 19

Goldstein, J., G. Siviglia, R. Hurst, L. Lenny and L. Reich, 'Group B erythrocytes enzymatically converted to Group O, survive normally in A, B and O individuals' in *Science*, 215, 1982, 168–170

Hollingham, R., 'Bloody beginnings', 'Affairs of the heart' and 'Dead man's hand' in *Blood and Guts: A History of Surgery*, BBC Books, London, 2008, 31–101, 103–155, 157–206

Moore, W., 'The Chaplain's neck' in *The Knife Man*, Bantam Press, London, 2005

Nuland, S.B., 'Medical science comes to America' in *Doctors: The Illustrated History of Medical Pioneers*, Black Dog & Leventhal Publishers Inc., New York, 2008, 366–401

Pain, S., 'One heart beating for two' in *New Scientist*, 28 July 2007, 50–51

Tucker, H., *Blood Work: A Tale of Medicine and Murder in the Scientific Revolution*, W.W. Norton & Co., London, 2011

HEARTACHES

Aldhous, P., 'Lab grown bladder shows big promise' in *New Scientist*, 8 April 2006, 10

Biever, C., 'Breathing life into artificial organs' in *New Scientist*, 3 February 2007, 24–25

British Heart Foundation, 'Open your heart' in *The New Review* in *The Observer*, 13 November 2011, 30

Burnie, D., 'The rise of modern surgery' and 'Technology and the body' in *Milestones in Medicine*, Reader's Digest Association Ltd, London, 1998, 65–85, 76–79

Churchill, W.S., *My Early Life: A Roving Commission*, Collins, London, 1980

Coghlan, A., 'How a foetus hides from its mother' in *New Scientist*, 3 November 2007, 16

Coghlan, A., 'Hybrid hearts for transplant' in *New Scientist*, 6 June 2009, 8–9

Coghlan, A., 'Stem cells for the masses' in *New Scientist*, 16 July 2011, 8–9

Editorial, 'Transplant hope as livers grown in the laboratory' in *The Observer*, 31 October 2010, 12

Editorial, 'First animal-to-human transplant' in *New Scientist*, 18 December 2010, 5

References

Editorial, 'Intestine grown from stem cells for the first time' in *New Scientist*, 18 December 2010, 18

Editorial, 'Womb transplant' in *New Scientist*, 18 June 2011, 7

Geddes, L., 'Organ transplants without lifelong drugs' in *New Scientist*, 23 February 2008, 9

Hollingham, R., 'Affairs of the heart' and 'Dead man's hand' in *Blood and Guts: A History of Surgery*, BBC Books, London, 2008, 103–155, 157–206

Ikenson, B., 'Artificial heart' [Kolff's patent] in *Patents: Ingenious Inventions*, Black Dog & Leventhal Publishers Inc., New York, 2004, 16–18

James, P., and N. Thorpe, 'False teeth and dentistry' in *Ancient Inventions*, Michael O'Mara Books Ltd, London, 1995, 33–36

Jones, G., and S. Sagee, 'Xenotransplantation [animal to human]: hope or delusion?' in *Biologist* 48, 2001, 129–132

Medawar, P.B., 'Tests by tissue culture methods on the nature of immunity to transplanted skin' in *Quarterly Journal of Microscopical Science*, 89, 1948, 239–252

Middleton, D., 'HLA: not just the scourge of transplant surgeons' in *Biologist*, 44, 1997, 322–325

Moore, W., 'The chimney sweep's teeth' in *The Knife Man*, Bantam Press, London, 2005, 151–179

Noble, J.W., 'Tooth Transplantation: a controversial story' in www.rcpsglasg.ac.uk/hdrg/home.html

Nuland, S.B., 'New hearts for old' in *Doctors: The Illustrated History of Medical Pioneers*, Black Dog & Leventhal Publishers Inc., New York, 2008, 436–467

O'Donaghue, K., and N. Fisk, 'Potential applications of stem cells' in *Biologist* 51 (3), 2004, 125–129

Pain, S., 'Can't bite, can't fight' in *New Scientist*, 10 March 2007, 50–51

Porter, R., 'Surgery' in *The Greatest Benefit to Mankind*, Fontana Press, London, 1999, 597–627

Power, H., 'Transplant rejection' in P. Tallack (ed.), *The Science Book*, Weidenfeld & Nicolson, London, 2001, 356–357

Roach, S., 'Whatever happened to the breakthroughs promised by stem cells?' in *The New Review* in *The Observer*, 11 August 2013, 19

Simmons, J.G., 'Ambroise Paré', 'Alexis Carrel' and 'Willem J. Kolff' in *Doctors and Discoveries*, Houghton Mifflin Co., Boston and New York, 2002, 119–122, 199–204, 275–279

Slezak, M., 'Get your spares at the organ farm' in *New Scientist*, 29 June 2013, 6–7

Thompson, A.W., and G.R.D. Catto, *Immunology of Renal Transplantation*, Edward Arnold, London, 1993

Van Dulken, S., 'The artificial heart' [Jarvik's patent] in *Inventing the 20th Century*, British Library, London, 2002, 172–173

THE GENIE IN THE GENE

Aldridge, S., 'Genetic engineering' and 'Human genome sequence' in P. Tallack (ed.) *The Science Book*, Weidenfeld & Nicolson, London, 2001, 436–437, 524–525

Bailey, A., 'Who was . . . Rosalind Franklin?' in *Biologist*, 50 (2), 2003, 92–93

Bragg, L., 'X-ray crystallography' in *Scientific American*, July 1968

Campbell N.A., and J.B. Reece, 'Mendel and the gene idea' and 'The molecular basis of inheritance' in *Biology*, 6th edn, Benjamin Cummings, San Francisco, 2002, 247–268, 287–301

Connor, S., 'Genetic odds that juries can count on' in *The Independent on Sunday*, 5 August 1990, 40–41

Dahl, L.F., 'X-ray diffraction' in S.P. Parker (ed.), *Concise Encyclopedia of Science and Technology*, 3rd edn, McGraw-Hill, New York, 1994, 2061–2063

Editorial, 'Deep down we're all faulty mutants' in *New Scientist*, 15 December 2012, 19

Granger, J., and D. Madden, 'The Polymerase chain reaction: Turning needles into haystacks' in *Biologist*, 40 (5), 1993, 197–200

Itzhaki, J.E., 'Hunting the genes for inherited diseases' in *Biologist* 43 (4), 1996, 173–175

Jones, H.D., and P.R. Wiley, 'The double helix 50 years on' in *Biologist*, 50 (2), 2003, 53–57

Macinnis, P., 'Discovering genetics', 'Discovering X-ray diffraction', 'Discovering the polymerase chain reaction' and 'Deducing the structure of DNA' in *100 Discoveries*, Pier 9, Murdoch Books UK Ltd, London, 2009, 176–178, 229–231, 246–247, 261–263

McKie, R., 'How the story of science's greatest breakthrough was almost never told' in *The Observer*, 12 September 2012, 25

Milton, J., 'Perfect technique' [interview with Kary Mullis] in *Biologist*, 59 (1), 2012, 27–28

More 4 TV (UK), *DNA, the Story of Life*, 23 February 2013

Parry, B., 'Sir Alec Jeffreys in conversation' in *Biologist*, 55 (1), 2008, 40–42

Ridley, M., 'Genes in inheritance' and 'Genetic fingerprinting' in P. Tallack (ed.), *The Science Book*, Weidenfeld & Nicolson, London, 2001, 264–265, 484–485

Rook, A. (ed.), 'Gregor Mendel' in *The Origins and Growth of Biology*, Penguin Books, Harmondsworth, 1964, 294–311

Simmonds, J.G., 'Oswald Avery' in *Doctors and Discoveries*, Houghton Mifflin Co., Boston and New York, 2002, 113–116

Solway, A., *Genetics in Medicine*, Franklin Watts, London, 2007

Van Dulken, S., 'Genetic fingerprinting' [Jeffreys's patent] in *Inventing the 20th Century*, British Library, London, 2002, 202–203

Vella, F., 'Gregor Mendel and his times' in A. Allison (ed.) *Penguin Science Survey B*, Penguin Books, Harmondsworth, 1966, 189–201

Wambaugh, J., *The Blooding*, Bantam Press, London, 1989

Watson, J.D., *The Double Helix*, Weidenfeld & Nicolson Ltd, London, 1968

Watson, J.D., and F.H. Crick, 'A structure for deoxyribose nucleic acid' in *Nature*, 171, 1953, 737

THE LATEST THINGS

Boucher, C., and C. Smith, *Extraordinary Origins of Everyday Things*, Reader's Digest Association Ltd, London, 2009

Bryson, B., 'The Kitchen' in *At Home*, Black Swan, London, 2010

Buchanan, N. (ed.), 'Everyday Miracles' in *How Is It Done?*, Reader's Digest Association Ltd, London, 1990, 9–30

Charlton, H. (ed.), *1000 Makers of the Twentieth Century*, The Sunday Times, London, 1991

References

Colquhoun, K., *Taste*, Bloomsbury Publishing plc, London, 1994

Dale, R., and J. Gray, *Edwardian Inventions*, Star Books, Allen & Unwin, London, 1979

De Bono, E. (ed.), 'Hearing aid', 'Lawnmower', 'Corset', 'Zipper', 'Vacuum cleaner' and 'Safety razor' in *Eureka!*, Thames & Hudson Ltd, London, 1974, 60–61, 110–111, 123–124, 131, 146–147, 147

Homer, T., 'Ears' and 'Eyes' in *The Book of Origins*, Portrait, Piatkus, London, 2007, 146–149

Hughes, K., *The Short Life and Long Times of Mrs Beeton*, Fourth Estate, London, 2005

Ikenson, B., 'Bar-code', 'Velcro®', 'Zipper', 'Lawnmower', 'Post-It notes' and 'Vacuum cleaner' in *Patents: Ingenious Inventions*, Black Dog & Leventhal Publishers Inc., New York, 2004, 71–73, 102–103, 103–104, 211–213, 220–221, 229–231

James, P., and N. Thorpe, 'Personal effects' in *Ancient Inventions*, Michael O'Mara Books Ltd, London, 1995, 242–297

Kerr-Jarrett, A., 'Life in the Victorian home' in *Life in the Victorian Age*, Reader's Digest Association Ltd, London, 1994, 37–49

Mrs Beeton, *The Book of Household Management. Also Sanitary, Medical & Legal memoranda; with a History of the Properties and Uses of All Things Connected with Home Life and Comfort* [1861] Chancellor Press, London, 1982

Oliver, H., *Cat Flaps and Mousetraps: The Origin of Objects in Our Daily Lives*, Metro Publishing, London, 2007

Pain, S., 'Don't forget your umbrella' in *New Scientist*, 30 May 2009, 42–43

Taylor, G.R., 'The coming of the contact lens', 'Lawn mower' and 'Razor' in *Inventions That Changed the World*, Reader's Digest Association Ltd, London, 1982, 76, 150, 225

Van Dulken, S., 'The safety razor' [Gillette's patent], 'The vacuum cleaner' [Booth], 'The zip fastener' [Sundback], 'Transparent adhesive tape' [Drew], 'The ballpoint pen' [Biro] and 'Bar codes' [Woodland and Silver], 'Velcro® fasteners' [De Mestral] and 'Post-It notes' [Silver] in *Inventing the 20th Century*, British Library, London, 2002, 30–31, 34–35, 58–59, 80–81, 106–107, 108–109, 144–145, 180–181

References

West, R.S., and M. Rickaby (eds), 'Home comforts' and 'The cleaning revolution' in *Inventions That Changed the World*, Reader's Digest Association Ltd, London, 1997, 16–17, 22–23

DIGITAL DOODAH

Austen, K., 'Out of the lab and into the streets' in *New Scientist*, 29 June 2013, 48–51

Burke, J., 'Faith in numbers' in *Connections*, Macmillan London Ltd, London, 1978, 81–113

Campbell, M., 'Power to the people' in *New Scientist*, 3 August 2013, 20

Editorial, 'Artificial symphony' in *New Scientist*, 7 July 2012, 7

Feldman, A., and P. Ford, 'Joseph-Marie Jacquard' and 'Charles Babbage' in *Scientists and Inventors*, Aldus Books, Middlesex, 1979, 82–83, 124–125

Graham-Cumming, J., 'Let's build a Babbage' in *New Scientist*, 18 December 2010, 26–27

Green, M., 'Errors of judgment' and 'Small wonders' in *The Nearly Men*, Tempus Publishing, Stroud, 2007, 215–271, 273–300

Hillis, W.D., 'Nuts and Bolts' in *The Pattern on the Stone: The Simple Ideas that Make Computers Work*, Basic Books, New York, 1998, 1–20

Holmes, R., 'Young scientists' in *The Age of Wonder*, HarperPress, London, 435–466

Hooper, R., 'One minute with Ada Lovelace. Extracts from her letters' in *New Scientist*, 13 October 2012, 29

Magnusson, M. (ed.), 'Computers – marvels in miniature' in *Reader's Digest Book of Facts*, Reader's Digest Association Ltd, London, 1994, 244–247

Mankiewicz, R., 'The computer' in P. Tallack, *The Science Book*, Weidenfeld & Nicolson, London, 2003, 340–341

Spangenburg, R., and D.K. Moser, 'Babbage, Lovelace and the first computers' in *The Age of Synthesis 1800–1895*, Facts on File Inc., New York, 62–63

Van Dulken, S., 'The computer' [Eckert and Mauchly patent] in

Inventing the 20th Century, British Library, London, 2002, 110–111

Webb, R., 'The hole story' in *New Scientist*, 19 November 2011, 47–49

West, R.S., and M. Rickaby (eds), 'Creating the computer and making it work' in *Inventions That Changed the World*, Reader's Digest Association Ltd, London, 1997, 40–41

Wilkinson, C., 'Tim Berners-Lee' and 'The Matrix' in *The Observer Book of Inventions*, Observer Books, London, 2008, 100, 101

COMING SOON

Energy, where art thou?

Barras, C., 'Will the Anaconda or the Oyster rule the waves?' in *New Scientist*, 6 March 2010, 18–19

Biever, C., 'Fusion star takes shape' in *New Scientist*, 7 April 2012, 22–23

Forshaw, J., 'Harnessing nuclear fusion' in *The Observer*, 16 September 2012

Giles, J., 'Staying power' in *New Scientist*, 5 January 2013, 28–31

Graham-Rowe, D., 'Is night falling on classic solar panels?' in *New Scientist*, 18 December 2010, 21

Hambling, D., 'Catching the sun' in *New Scientist*, 13 August 2011, 36–39

Hodson, H., 'Greening the grid' in *New Scientist*, 2 February 2013, 20

Hodson, H., 'Wave power reborn' in *New Scientist*, 25 May 2013, 22

Hodson, H., 'Out of thin air' in *New Scientist*, 3 August 2013, 22

Knight, H., 'Biofuel turns to a healthier diet' in *New Scientist*, 29 May 2010, 22–23

Knight, H., 'The power of the cool' in *New Scientist*, 26 February 2011, 21

McKenna, P., 'One minute with Raymond Mabus' in *New Scientist*, 7 May 2011, 29

McKenna, P. 'Riding the wave' in *New Scientist*, 27 August 2011, 17–18

McKenna, P., 'Melt buildings to save fuel' in *New Scientist*, 7 January 2012, 17–18

References

Marks, P., 'Supercool wind power' in *New Scientist*, 19 January 2013, 19

Marshall, M., 'Panel price crash could spark solar revolution' in *New Scientist*, 14 February 2012, 12

Pierce, F., 'Rising tide' in *New Scientist*, 17 September 2011, 48–51

Platt, R., 'Cut the Bluster' in *New Scientist*, 19 January 2013, 26–27

Sample, I., 'Use vats of bacteria to make jet fuel' in *The New Review* in *The Observer*, 27 February 2011, 22

Various, 'Renewable energy' in *New Scientist*, 11 October 2008, 28–41

Who's in charge?

Biever, C., 'I robot' in *New Scientist*, 18 May 2013, 40–41

Biever, C., 'Machines come to life' in *New Scientist*, 10 August 2013, 8–9

Carroll, C., 'Us and them' in *National Geographic*, August 2011, 66–85, 2013, 9

Cooper, K., 'Toot toot: road is cleared for driverless cars' in *The Sunday Times*, 12 May 2013

Editorial, 'Blind man drives Google car' in *New Scientist*, 7 April 2012, 23

Giles, J., 'What's next for Watson?' in *New Scientist*, 19 February 2011, 6–7

Graham-Cumming, J., 'Alan Turing's legacy' in *New Scientist*, 2 June 2012, i–viii

Marks, P., 'Hands off the wheel' in *New Scientist*, 31 March 2012, 19–20

Marks, P., 'Now it's Grand Hack auto' in *New Scientist*, 20 July 2013, 20

Norvig, P., 'Artificial intelligence' in *New Scientist*, 3 November 2012 i–viii

Russell, S., and P. Norvig, *Artificial Intelligence: A Modern Approach*, 3rd edn, Prentice Hall, New York, 2009

Smith, B.W., 'Who's the driver?' in *New Scientist*, 22–29 December 2012, 34–35

Turing A., 'Computing machinery and intelligence' in *Mind*, 59, 1950, 433

Bionic man

BBC2 TV, *Horizon: The Hunt for Artificial Intelligence*, 3 April 2013

Boyd, J., 'Dress for action with bionic suit' in *New Scientist*, 9 April 2005, 19

Butler, S., *Erewhon* [1872], 10th rev. edn, A.C. Fifield, London, 1913

Channel 4 TV (UK), *How to Build a Bionic Man*, 7 February 2013

Coghlan, A., 'Brain-controlled robot arm could beat paralysis' in *New Scientist*, 19 May 2012, 10

De Sautoy, M., 'AI robot: evolution of the machines that learn for themselves' in *The New Review* in *The Observer*, 1 April 2012, 24–25

Editorial, 'Need a jaw bone? Print one' in *New Scientist*, 11 February 2012, 7

Geddes, L., 'A cyborg is born' in *New Scientist*, 24 September 2011, 25

Gilhooly, R., 'Full metal jacket' in *New Scientist*, 21 April 2012, 19–20

Hodson, H., 'Need a hand? Have two' in *New Scientist*, 27 October 2012, 18

Jabr, F., 'Mind-controlled robotic arm to help amputees' in *New Scientist*, 30 April 2011, 11

Thomson, H., 'Get your move on' in *New Scientist*, 8 June 2013, 19

SCIENCE FICTIONS?

Raising the dead

Adams, T., 'This man can bring you back from the dead' in *The New Review* in *The Observer*, 7 April 2013, 18–19

Albin-Dyer, B., 'Cryonics: the techno funeral' in *Don't Drop the Coffin*, Hodder & Stoughton, London, 2002, 255–280

BBC2 TV, *Horizon: Back from the Dead*, 27 October 2010

Boia, L., 'Freezing, clones and robots' in *Forever Young*, Reaktion Books, London, 192–196

Bondeson, J., 'Apparent death and premature burial' in *A Cabinet of Medical Curiosities*, I.B. Tauris Publishers, London, 1997, 96–121

Collins, P., 'Poe's cure for death' in *New Scientist*, 13 January 2007

Fong, K., *Extremes*, Hodder & Stoughton, London, 2013

Geddes, L., 'When death isn't what it seems' in *New Scientist*, 8 October 2011, 8–9

Hughes, J., 'Vital signs' in *New Scientist*, 13 October 2007, 50–51

Moore, W., 'The Chaplain's neck' in *The Knife Man*, Bantam Press, London, 2005, 266–294

Oddy, J., 'Cool Customers' in *The Independent on Sunday*, 14 July 2002, 18–22

Parnia, S., *The Lazarus Effect: The Science that is Rewriting the Boundaries between Life and Death*, Ebury Press, London, 2013

Sawyer, A., 'Future Foretold' in *New Scientist*, 12 May 2011

Teresi, D., 'Resurrection man' (interview with Sam Parnia) in *New Scientist*, 9, March 2013, 32–33

Van Dulken, S., 'The deliverance coffin' in *Inventing the 19th Century*, British Library, London, 2001, 70–71

Never say die

Appleyard, B., 'The woman who won't die' in *The Sunday Times*, March 16 1997, 48–50

Coghlan, A., 'Red wine's anti-ageing ingredient does it again' in *New Scientist*, 11 February 2006, 12

De Langue, C., 'It was like finding a thing that shouldn't be' in *The New Review* in *The Observer*, 17 March 2013, 22

Editorial, 'How to live long and prosper' in *New Scientist*, 25 November 2006, 19

Editorial, 'DNA trick throws ageing into reverse' in *New Scientist*, 4 December 2010, 19

Editorial, 'Eunuchs provide clues to longer life' in *New Scientist*, 29 September 2012, 14

Egan, D., 'We're going to live forever' in *New Scientist*, 10 October 2007, 46

References

Haycock, D.B., *Mortal Coil: A Short History of Living Longer*, Yale University Press, New Haven and London, 2008

Heaven, D., 'Master key opens door to longer life' in *New Scientist*, 4 May 2013, 8–9

Hooper, R., 'Unravelling the secrets of ageing' in *New Scientist*, 30 July 2005, 11

Klerkx, G., 'The immortals' club' in *New Scientist*, 9 April 2005, 38–41

Lawton, G., 'The incredibles' in *New Scientist*, 13 May 2006, 32–38

Motluck, A., 'How long have you got?' in *New Scientist*, 10 December 2011, 46–49

Sohal, R.S., and R. Weindruch, 'Oxidated stress, calorific restriction and ageing' in *Science*, 273, 1996, 1581–1588

Creating life

Adams, T., 'Craig Venter. The first lord of the laboratory' in *The Observer*, 23 May 2010, 26

BBC2 TV, *Horizon: Playing God*, 17 January 2012

Campbell, N.A., and J.B. Reece, 'Carbon and the molecular diversity of life' in *Biology*, 6th edn, Benjamin Cummins, San Francisco, 2002, 52–61

Darwin, C., *The Descent of Man*, John Murray, London, 1871

Editorial, 'Why life on Earth was a sure thing' in *New Scientist*, 16 December 2006, 16

Editorial, 'Volcanoes' vital contribution to the primordial melting pot' in *New Scientist*, 25 October 2008, 14

Editorial, 'DNA transistor heralds living computers' in *New Scientist*, 6 April 2013, 18

Hamzelou, J., 'Mice born from eggs built in the lab' in *New Scientist*, 13 October 2012, 15

Jabr, F., 'Viable mouse sperm grown from scratch' in *New Scientist*, 26 March 2011, 18

Marshall, M., 'In the beginning was Pac-Man' in *New Scientist*, 5 March 2011, 8–9

Marshall, M., 'Synthetic version of DNA created' in *New Scientist*, 20 April 2012, 10

References

Ravilious, K., 'With the right recipe, early life's a cinch' in *New Scientist*, 16 May 2009, 13

Sanderson, K., 'The life factory' in *New Scientist*, 29 January 2011, 32–35

Tallack, P., 'Synthesis of urea' in *The Science Book*, Weidenfeld & Nicolson, London, 2003, 142–143

Witham, L., 'Life's origin' in *By Design*, Encounter Books, San Francisco, 2003, 95–112

INDEX

427

Index

Index

Kelvinator Inc. 344–5
Kemmler, William 165
Kenall, Captain 196–7
keys 334
kidneys
    dialysis 300–301
    transplants 309–10,
        311
Kiesler, Hedwig 86–7
Kilby, Jack 354–5
kilns 8
Kinecolor 228
Kinetoscope 222
kites 109–10, 124, 154
Klaw, Pauline 330
Knight, J.P. 135
knives 3, 4
Knossos palace lavatory
    274
Koch, Robert 264–5, 266,
    267
Kodachrome 219
Kodak 217–18, 219
Kolff, Willem 300–301,
    312–13
Kruesi, John 208–9
Kühnke, Karl 102

Labrousse, Jeanne 129
Laënnec, René 288, 289
Lalande, Joseph 129–30
Landsteiner, Karl 298–9
Langen, Eugen 42
Langevin, Paul 202
Lassie 226
lassoes 6–7
laughing gas 282–3
Lauste, Eugène 227
Lauterbur, Paul 295
Lavassor, Émile 45
lavatories 274–80
lawnmowers 340–41
Lawrence, Florence 225
Lazy Bones (remote
    control) 240
Le Neve, Ethel 196–7
Le Prince, Augustin
    222–3
Leach, Bernard 8–9
lead additive, to petrol
    345
Leden, Judy 122
Leeuwenhoek, Antonie van
    251–2
LEGO 56

Leibniz, Gottfried 347
Lejust, Léon 146
Lenoir, Étienne 41, 46
Lenormand, Louis-
    Sébastien 128
lenses
    contact 329
    lighthouse 143–4
    microscope 251, 260
    spectacle 245
    telescope 245–6, 247
Leonardo da Vinci 14, 54,
    74, 122, 127–8, 329
Leonardo da Vinci (battle-
    ship) 98
Lethbridge, John 93–4
Letheon 284
levers 14
Lewis, John 313
Lieben, Robert von 231
life-saving pumps 374
life creation 381–5
life jackets 145–7
lifespan extension 376–81
light bulbs 160–63, 166
lighthouses 140–44
lightning conductors
    152–5
lightships 140
Lilienthal, Otto 121–3
Lillehei, Walter 302, 313,
    314
Lillehci–DeWall Bubble
    Oxygenator 303
Lincoln, Abraham 215
Linde, Karl von 344
Lippershey, Hans 246
Lister, Agnes 261
Lister, Joseph 260–64
Listerine 264
Liston, Robert 281, 285
lithium 362
Liverpool to Manchester
    Railway 31–3
Livingston, Robert 65
locks 334
Locomotives on Highways
    ('Red Flag') Act 40–41
Lodge, Oliver 191–2, 193,
    194, 195, 197
London 40, 285
    drinking water 276,
        277
    Great Exhibition
        (1851) 276

Great Midlands Hotel
    274
Royal Society *see* Royal
    Society
sewage disposal 276–7,
    278–9
Long, Crawford 284
longevity 376–81
Longmore, Donald 316
longships 61
looms, programmable 348
Lorena, Guglielmo de 90
lost wax casting 9
Loud, John 332
loudspeakers 231
Louis I of Spain 255
Louis XIV 141
Louis XV 255, 273
Lovelace, Ada Byron,
    Countess of 350–51,
    357
Lovell, Sir Bernard 203
Lucretius 153
Lumière brothers (Auguste
    and Louis) 218, 223
Lunardi, Vicenzo 115
lung machines 303
Luther, Martin 245
Lynch, Bernard 132

McAdam, John 37
macadamisation 37
Macaulay, Thomas
    Babington 255
Macintosh, Robert 287
Maddox, Richard 216
Maes Howe, Orkney 12
Magnetic Resonance
    Imaging (MRI) 295
Magnetophone 211
magnetrons 204–5
magnifying glasses 250
Mahan, Steve 363
Malthus, Thomas Robert
    259
Manchester 24
Mannes, Leopold 219
Mansfield, Peter 295
Marchi, Francesco de 90
Marconi, Guglielmo
    192–7, 232–3, 331
Marconi Wireless
    Telegraph 193
Marcus Aurelius 272
Marey, Étienne-Jules 222

Index

Marie-Antoinette 113
Marie, David 148
Martin Baker (company)
    131–2
Mary Rose 96
Maskelyne, Edmund 279
Maskelyne, Nevil 194, 249
masking tape 332
mattocks 9
Mauchly, John 352
Maury, M.F. 175
Maxim, Hiram 44
Maxwell, James Clerk 218
May, Wesley 131
Medawar, Peter 310
medicine/medical practice
    254–70
  bionics 368–71
  and death reversal
    374–5
  diagnostic tools *see*
    CAT scanner;
    Magnetic Resonance
    Imaging; PETT
    scanner; stetho-
    scopes; ultrasound
    scanning; X-rays
  gene therapy 326–7
  kidney dialysis
    300–301
  stem cell engineering
    317–18
  surgical practice *see*
    surgical practice
Melba, Dame Nellie 233
Mendel, Gregor 319–21
Mercedes 46
Messenger, Thomas 330
metal detectors 188
Meucci, Antonio 182–3
MGM 226
miasma theory 254, 267
Mickey Mouse 226
microbes 251–3, 262,
    264–6, 267–70, 358–9
  see also bacteria
microfilming 117
microphones 186, 201,
    227, 234
microprocessors 355
microscopes 250–52, 260,
    264
microwave ovens 204–5
Middleton, Hugh 276
Midgley, Thomas Jnr 345

Miles Martin Pen
    Company 332
Miller, Patrick 62–3
Miller, Stanley 382
mills 17–20
Milner, Thomas 335
mirrors, telescope 247–8,
    249
mitochondrial Eve 1
mobile phones 190
Monitor 72
monoliths 11
montage 225
Montagu, Lady Mary
    Wortley 255
Montgolfier, Étienne 111,
    112, 113–14
Montgolfier, Joseph 111,
    112, 113–14
Montrose 196–7
moon-landing 240
Moore, Gordon 355
Morgan, Garrett 136
Morse, Samuel 171–2, 173
Morse code 171–2, 193,
    194, 195, 208–9, 230
Morton, William 283–4
motion pictures 220–29
motor cars *see* cars
Motorwagen 43
movies *see* films (movies)
Movietone newsreels 229
MRI (Magnetic Resonance
    Imaging) 295
Mullis, Kary 324
Munter, Carl 344
Murray, George 169
Murray, Joseph 311
Mussolini, Benito 90
mutoscopes 220
Muybridge, Eadweard
    220–22

Nakauchi, Hiromitsu 317
Napoleon I 61–2, 79, 80,
    156
Nassau 129
National Anti-Vaccination
    League 258
National Geographic
    Society 188
Nautilus 78–9
Neanderthals 4–6
needles 5, 208, 255, 257,
    261

Nelson, Horatio, 1st
    Viscount 63, 80
Nero 328
Neufeldt, Hans 102
New Grange, Ireland 12
Newcomen, Thomas,
    steam engine 21–2
Newshaw, Richard 148
Newton, Isaac 247–8, 328
Niagara, USS 176–7
Nicholas II 325
Nickelodeons 224
Nickerson, William 336
Niépce, Nicéphore 214
Nipkow, Paul 235
nitrous oxide 282–3
Norberg, Jonas 143
Northumbrian,
    Stephenson's 29–30
Noyce, Robert Norton
    355
nuclear fusion 362–3

O'Brien, Beatrice 197
Obry, Ludwig 85–6
Octavius, Prince 256
oil 358–9
O'Neill, Peggy 237–8
opium 281
Oppenheimer, Benjamin
    148–9
optical lenses *see* lenses
organ transplants 309–18
  rejection 310–18
Orwell, George 241, 372
Otophone 331
Otto, Nikolas 41–2
Ovid 107–8

pacemakers 312
Painter, William 335
Pan Xiting 377
Panhard, René 45
Papin, Denis 20–21
parachutes 127–31, 148–9
Paré, Ambroise 296–7,
    306, 309
Paris, J.A. 220
Paris Exposition (1900)
    223
Parker Pen Company 332
Parsons, Charles 72–3
Pascal, Blaise 347
Pask, Edgar 146–7
Pasley, Charles 97–8

434

Index

vaccination 255–6, 257–9, 265–6
vaccines 265–6
vacuum cleaners 342–3
vacuum tubes 231
Vail, Alfred 171–2
Vail Ironworks 171
valves
  artificial heart valves 311–12
  radio 231–2, 234, 351–2
Vaucanson, Jacques de 347–8
Velcro 337
Venter, Craig 383–4
Vermeer, Johannes 252
Verne, Jules 372
Vespasiano da Bisticci 207
VHF transmissions 233, 239
Victoria, Queen 177, 215, 229, 258, 276, 286–7
Victoria, HMS 71
Victory, HMS 70
vinyl discs 211
Virginia 71–2
Vodaphone 189
Volta, Alessandro 155–6, 330
Voltaire 154
Volvo 133–4
von Neumann, John 352
Vulcanite 55

Walkman 212
Wall-E 365
Wallace, Alfred Russel 258–9
Walpole, Sir Robert 15–16
Warner Brothers 226
Warren, John C. 283
warships 61, 70–72, 86, 88, 98–100

Washington, George 77, 305–6
Washkansky, Louis 315
water power 16–20, 360–61
Waterman, Lewis 331–2
waterwheels 17
Watson (computer) 366–7
Watson, James 292, 321, 322–3
Watson, Thomas 184, 187, 188
Watson, Sir William 173
Watson-Watt, Robert (later Sir Robert) 199–200, 205
Watt, James 22–3, 26, 36–7, 40, 62, 65
weapons
  prehistoric 2, 3, 4, 5, 10
  torpedoes 85–8
weaving/weavers 9, 19–20, 337, 348
  see also cotton spinning
Webb, Jane 372
Welles, Orson 234, 342
Wells, H.G. 125
  War of the Worlds radio adaptation 234
Wells, Horace 282–3
Western Union 172, 183, 185
Westinghouse, George 164, 165, 166–7, 233, 239
Wheatstone, Charles 170–71, 174
wheels 52
  paddle 60–61, 67, 76
  potter's 8–9
  water 17
Whitehead, Robert 85, 86
Wichterle, Otto 329

Wilhelm II 119
Wilhelm Röntgen 289–90, 292
Wilkins, John 76
Wilkins, Maurice 321–2
Wilkinson, John 'Iron-Mad' 23
William, Duke of Gloucester 255
Williams, Freddie 352
Williamson, James 229
Willis, Thomas 329–30
Wilson, Arthur, Rear Admiral 84
Wilson, Taylor 362
wind turbines/farms 360, 361
Windsor Castle 276
Winstanley, Henry 141–2
wire/cable insulation 169, 171, 173, 174, 211
wireless waves 87
Woodland, Norman 333
Woodward, Henry 161
Woodward, Joseph 201–2
Wordsworth, William 11
World Wide Web 356–7
Wren, Christopher 141, 298
Wright, Orville 123
Wright, Wilbur 120, 123–4, 125
writing 206–8

X-rays 289–92, 314, 357
XCR (robot) 367

Yale, Linus Jnr 334

Zeppelin, Ferdinand 119
Zeppelins 119–20
zip fasteners 336
Zworykin, Vladimir 239